# Invariants of Boundary Link Cobordism

of the
American Mathematical Society

Number 784

# Invariants of Boundary Link Cobordism

Desmond Sheiham

September 2003 • Volume 165 • Number 784 (first of 4 numbers) • ISSN 0065-9266

**American Mathematical Society**
Providence, Rhode Island

2000 *Mathematics Subject Classification.*
Primary 18F25, 57Q45, 57Q60, 16G20.

---

**Library of Congress Cataloging-in-Publication Data**

Sheiham, Desmond, 1974–
  Invariants of boundary link cobordism / Desmond Sheiham.
    p. cm. — (Memoirs of the American Mathematical Society, ISSN 0065-9266 ; no. 784)
  "Volume 165, number 784 (first of 4 numbers)."
  Includes bibliographical references and index.
  ISBN 0-8218-3340-5 (alk. paper)
  1. K-theory. 2. Knot theory. 3. Cobordism theory. I. Title. II. Series.

QA3.A57  no. 784
[QA612.33]
510 s—dc21
[512′.55]                                                                             2003051903

---

# Memoirs of the American Mathematical Society

This journal is devoted entirely to research in pure and applied mathematics.

**Subscription information.** The 2003 subscription begins with volume 161 and consists of six mailings, each containing one or more numbers. Subscription prices for 2003 are $555 list, $444 institutional member. A late charge of 10% of the subscription price will be imposed on orders received from nonmembers after January 1 of the subscription year. Subscribers outside the United States and India must pay a postage surcharge of $31; subscribers in India must pay a postage surcharge of $43. Expedited delivery to destinations in North America $35; elsewhere $130. Each number may be ordered separately; *please specify number* when ordering an individual number. For prices and titles of recently released numbers, see the New Publications sections of the *Notices of the American Mathematical Society*.

**Back number information.** For back issues see the *AMS Catalog of Publications*.

Subscriptions and orders should be addressed to the American Mathematical Society, P. O. Box 845904, Boston, MA 02284-5904, USA. *All orders must be accompanied by payment.* Other correspondence should be addressed to 201 Charles Street, Providence, RI 02904-2294, USA.

**Copying and reprinting.** Individual readers of this publication, and nonprofit libraries acting for them, are permitted to make fair use of the material, such as to copy a chapter for use in teaching or research. Permission is granted to quote brief passages from this publication in reviews, provided the customary acknowledgment of the source is given.

Republication, systematic copying, or multiple reproduction of any material in this publication is permitted only under license from the American Mathematical Society. Requests for such permission should be addressed to the Acquisitions Department, American Mathematical Society, 201 Charles Street, Providence, Rhode Island 02904-2294, USA. Requests can also be made by e-mail to reprint-permission@ams.org.

---

*Memoirs of the American Mathematical Society* is published bimonthly (each volume consisting usually of more than one number) by the American Mathematical Society at 201 Charles Street, Providence, RI 02904-2294, USA. Periodicals postage paid at Providence, RI. Postmaster: Send address changes to Memoirs, American Mathematical Society, 201 Charles Street, Providence, RI 02904-2294, USA.

© 2003 by the American Mathematical Society. All rights reserved.
This publication is indexed in *Science Citation Index*®, *SciSearch*®, *Research Alert*®, *CompuMath Citation Index*®, *Current Contents*®/*Physical, Chemical & Earth Sciences*.
Printed in the United States of America.

∞ The paper used in this book is acid-free and falls within the guidelines established to ensure permanence and durability.
Visit the AMS home page at http://www.ams.org/

10 9 8 7 6 5 4 3 2 1      08 07 06 05 04 03

# Contents

| | | |
|---|---|---|
| Acknowledgements | | ix |
| Chapter 1. Introduction | | 1 |
| 1. Isotopy and Cobordism; the Link | | 2 |
| 2. Knot Theory | | 3 |
| 3. Boundary Links | | 4 |
| 4. Surgery Obstructions | | 7 |
| 5. Algebraic $F_\mu$-link Cobordism | | 13 |
| Chapter 2. Main Results | | 17 |
| 1. Signatures | | 18 |
| 2. Number Theory | | 21 |
| 3. Defining Invariants | | 22 |
| 4. Chapter Summaries | | 27 |
| Chapter 3. Preliminaries | | 29 |
| 1. Representations | | 29 |
| 2. Grothendieck Groups | | 31 |
| 3. Witt Groups | | 32 |
| 4. Seifert Forms | | 37 |
| Chapter 4. Morita Equivalence | | 41 |
| 1. Linear Morita Equivalence | | 42 |
| 2. Hermitian Morita equivalence | | 43 |
| Chapter 5. Devissage | | 47 |
| Chapter 6. Varieties of Representations | | 51 |
| 1. Existence of Simple Representations | | 52 |
| 2. Semisimple Representations of Quivers | | 53 |
| 3. Self-Dual Representations | | 54 |
| Chapter 7. Generalizing Pfister's Theorem | | 59 |
| 1. Artin-Schreier Theory | | 59 |
| 2. Frobenius reciprocity | | 61 |
| 3. Proof of Theorem 7.7 | | 62 |
| Chapter 8. Characters | | 63 |
| 1. Artin algebras | | 63 |

|   |   |   |
|---|---|--:|
| | 2. Independence of Characters | 64 |
| Chapter 9. | Detecting Rationality and Integrality | 67 |
| | 1. Rationality | 67 |
| | 2. Integrality | 72 |
| Chapter 10. | Representation Varieties: Two Examples | 75 |
| | 1. Dimension $(1,1)$ | 75 |
| | 2. Dimension $(2,1)$ | 77 |
| Chapter 11. | Number Theory Invariants | 81 |
| | 1. Division Algebras over $\mathbb{Q}$ | 82 |
| | 2. Witt Invariants | 84 |
| | 3. Localization Exact Sequence | 88 |
| Chapter 12. | All Division Algebras Occur | 91 |
| | 1. Proof of Proposition 12.1 | 92 |
| | 2. Proof of Proposition 12.2 | 93 |
| | 3. Computation of $C_{2q-1}(F_\mu)$ up to Isomorphism | 96 |
| Appendix I. | Primitive Element Theorems | 99 |
| Appendix II. | Hermitian Categories | 101 |
| | 1. Functors | 101 |
| | 2. Duality Preserving Functors | 102 |
| Bibliography | | 105 |
| Index | | 109 |

## Abstract

An $n$-dimensional $\mu$-component boundary link is a codimension 2 embedding of spheres

$$L = \bigsqcup_\mu S^n \subset S^{n+2}$$

such that there exist $\mu$ disjoint oriented embedded $(n+1)$-manifolds which span the components of $L$. An $F_\mu$-link is a boundary link together with a cobordism class of such spanning manifolds.

The $F_\mu$-link cobordism group $C_n(F_\mu)$ is known to be trivial when $n$ is even but not finitely generated when $n$ is odd. Our main result is an algorithm to decide whether two odd-dimensional $F_\mu$-links represent the same cobordism class in $C_{2q-1}(F_\mu)$ assuming $q > 1$. We proceed to compute the isomorphism class of $C_{2q-1}(F_\mu)$, generalizing Levine's computation of the knot cobordism group $C_{2q-1}(F_1)$.

Our starting point is the algebraic formulation of Levine, Ko and Mio who identify $C_{2q-1}(F_\mu)$ with a surgery obstruction group, the Witt group $G^{(-1)^q,\mu}(\mathbb{Z})$ of $\mu$-component Seifert matrices. We obtain a complete set of torsion-free invariants by passing from integer coefficients to complex coefficients and by applying the algebraic machinery of Quebbemann, Scharlau and Schulte. Signatures correspond to 'algebraically integral' simple self-dual representations of a certain quiver (directed graph with loops). These representations, in turn, correspond to algebraic integers on an infinite disjoint union of real affine varieties.

To distinguish torsion classes, we consider rational coefficients in place of complex coefficients, expressing $G^{(-1)^q,\mu}(\mathbb{Q})$ as an infinite direct sum of Witt groups of finite-dimensional division $\mathbb{Q}$-algebras with involution. The Witt group of every such algebra appears as a summand infinitely often.

The theory of symmetric and hermitian forms over these division algebras is well-developed. There are five classes of algebras to be considered; complete Witt invariants are available for four classes, those for which the local-global principle applies. An algebra in the fifth class, namely a quaternion algebra with non-standard involution, requires an additional Witt invariant which is defined if all the local invariants vanish.

---

2000 *Mathematics Subject Classification.* 18F25, 57Q45, 57Q60, 16G20.

# Acknowledgements

I am very grateful to my PhD adviser, Andrew Ranicki, for the energy, ideas, encouragement and copious enthusiasm he has shared with me throughout my time as a graduate student in Edinburgh. I would also like to thank Michael Farber for valuable suggestions at an early stage of this thesis project.

Among many others with whom I have enjoyed interesting and useful mathematical conversations, I would particularly like to mention Richard Hill, David Lewis, Kent Orr and Peter Teichner who made it possible for me to spend time at their respective universities and were generous to me with their time and expertise.

CHAPTER 1

# Introduction

The classification problem which we address in this volume concerns $n$-dimensional spheres $S^n$ knotted and linked inside an $(n+2)$-dimensional space. This $(n+2)$-dimensional space could be $\mathbb{R}^{n+2}$, but following standard convention let us assume that it is the one point compactification, $S^{n+2}$.

We focus on boundary links, links whose components are the boundaries of *disjoint* $(n+1)$-dimensional manifolds inside $S^{n+2}$. In fact, our basic object of study is the $F_\mu$-link, a refinement of the ($\mu$-component) boundary link whose definition involves the free group $F_\mu$ on $\mu$ non-commuting generators; see definition 1.5 below.

Whilst all three notions - 'link', 'boundary link' and '$F_\mu$-link' - are generalizations of 'knot', the theory of $F_\mu$-links is most directly analogous to knot theory.

The aim of the present work is to provide the means to calculate whether two arbitrary $F_\mu$-links are 'the same' or 'different', up to an equivalence relation known as cobordism[1] [**29**]. In the definition of cobordism one regards the ambient sphere $S^{n+2}$ as the boundary of an $(n+3)$-dimensional disk, $D^{n+3}$. To say that a knot, for example, is cobordant to the trivial knot is *not* to say that one can untie the knot in $S^{n+2}$, but that one can untie it in $D^{n+3}$ by contracting it concentrically through some $(n+1)$-dimensional disk.

If $n \geq 2$ then the set $C_n(F_\mu)$ of cobordism classes of $F_\mu$-links is an abelian group; one adds two links by 'ambient connected sum', joining corresponding components of the two links with narrow tubes. The secondary aim of this work is to compute the isomorphism class of $C_n(F_\mu)$.

A detailed computation of the knot cobordism group $C_n(F_1)$ was achieved in the 1960's and 1970's for all $n \geq 2$; only the case $n = 1$ remains an open problem. The even-dimensional groups turned out to be trivial [**41**], whereas the odd-dimensional groups are not even finitely generated.

The odd-dimensional computation emerged in two stages. Firstly, surgery methods were used to identify $C_n(F_1)$ with a group defined in purely algebraic terms [**56, 11, 38, 39, 76, 77, 83, 95**]. Secondly, numerical invariants powerful enough to distinguish all the odd-dimensional knot cobordism classes were defined and the cobordism group was computed [**69, 55, 42, 97**].

---

Received by the editor October 24th 2001.
[1]Some authors prefer the synonym 'concordance' to 'cobordism'.

The first stage of the $F_\mu$-link cobordism computation, reduction to algebra, has also been achieved. The even-dimensional groups are trivial while the odd-dimensional groups are 'even larger' than in knot theory. This first chapter is a short exposition of the reduction to algebra of knot cobordism and $F_\mu$-link cobordism. We recall the different notions of link cobordism and state three (equivalent) identifications of $C_n(F_\mu)$ with surgery obstruction groups.

Our main results, which we state in chapter 2, concern the second stage of the $F_\mu$-link cobordism problem. We define an algorithmic procedure to decide whether or not two $F_\mu$-links are cobordant (assuming $q > 1$) and proceed to compute the isomorphism class of $C_{2q-1}(F_\mu)$. The definitions of our invariants also apply when $q = 1$ but they are far from a complete set in this case.

Previous odd-dimensional link cobordism signatures include those of S.Cappell and J.Shaneson [**12**, p46] and of J.Levine [**58, 59**] who obtained some $F_\mu$-link signatures via jumps in the $\eta$-invariants associated to unitary representations $F_\mu \to U(m)$ of the free group.

We employ quite different methods taking as a starting point the algebraic formulation of $C_{2q-1}(F_\mu)$ in terms of 'Seifert matrices' (sections 4.2 and 5.2 below).

## 1. Isotopy and Cobordism; the Link

Let us first explain what is meant by cobordism of links. A link is an embedding of disjoint $n$-dimensional spheres[2] in an $(n+2)$-dimensional sphere:

$$L \cong \overbrace{S^n \sqcup \cdots \sqcup S^n}^{\mu} \subset S^{n+2} .$$

Each component of a link may be knotted; indeed, a 1-component link is called a *knot*.

Two links are called isotopic and are usually considered to be 'the same' if one of the links can be transformed into the other through embeddings:

DEFINITION 1.1. Links $L^0$ and $L^1$ are *isotopic* if they can be joined[3] in $S^{n+2} \times [0,1]$ by an embedding

$$LI \cong (S^n \sqcup \cdots \sqcup S^n) \times [0,1] \subset S^{n+2} \times [0,1]$$

---

[2] More precisely, a link will be an embedding of disjoint smooth manifolds $\Sigma_1 \sqcup \cdots \sqcup \Sigma_\mu \hookrightarrow S^{n+2}$ such that each component $\Sigma_i$ is homeomorphic to $S^n$. Alternatively, one can interpret everything in the piecewise linear category, if one assumes that links, isotopies and cobordisms are locally flat PL embeddings. If one were to consider locally flat topological embeddings the theory could differ only when $n = 3$ or $4$. (cf Novikov [**74**, Theorem 6] and Cappell and Shaneson [**10**]).

[3] More precisely it is required that there exists a smooth oriented submanifold $LI$ of $S^{n+2} \times I$ such that $LI$ is homeomorphic to $(S^n \sqcup \cdots \sqcup S^n) \times [0,1]$ and meets $S^{n+2} \times \{0\}$ and $S^{n+2} \times \{1\}$ transversely at $L^0$ and $L^1$ respectively.

such that, for each $i \in [0,1]$,
$$(S^n \sqcup \cdots \sqcup S^n) \times \{i\} \subset S^{n+2} \times \{i\}.$$
A link $L$ is in fact isotopic to the trivial link if and only if $L$ is the boundary of $\mu$ disjoint disks embedded in $S^{n+2}$ (see for example Hirsch [37, Thm8.1.5]).

Cobordism is a weaker equivalence relation first defined in the context of classical knots $S^1 \subset S^3$ by Fox and Milnor [29]. One merely omits the condition that the embedding should be 'level-preserving':

DEFINITION 1.2. Links $L^0$ and $L^1$ are *cobordant* if they may be joined[3] by an embedding
$$LI \cong (S^n \sqcup \cdots \sqcup S^n) \times [0,1] \subset S^{n+2} \times [0,1].$$
The set of $n$-dimensional $\mu$-component link cobordism classes is denoted $C(n,\mu)$.

A cobordism between $L$ and the trivial link can be 'capped' at the trivial end by $\mu$ disjoint disks $D^{n+1} \subset S^{n+2}$. Thus a link is cobordant to the trivial link if and only if it bounds $\mu$ disjoint disks embedded in $D^{n+3}$:

(1)
$$\begin{array}{ccc} S^n \sqcup \cdots \sqcup S^n & \subset & S^{n+2} \\ \cap & & \cap \\ D^{n+1} \sqcup \cdots \sqcup D^{n+1} & \subset & D^{n+3} \end{array}$$

A link which is cobordant to a trivial link is called *null-cobordant* or *slice*.

The general problem of classifying knots and links up to isotopy appears very difficult, although much is known about certain restricted classes. On the other hand a detailed classification of knot cobordism $C(n,1)$ was achieved in the 1960's and 1970's for all $n \geq 2$. For $\mu \geq 2$ the computation of $C(n,\mu)$ is at the time of writing an important open problem.

## 2. Knot Theory

Before introducing boundary links, let us review a little high-dimensional knot theory. To the uninitiated reader we recommend Kervaire and Weber [40] and the (low-dimensional) books of Lickorish [65] and Rolfsen [87]. Further expositions in high dimension include Farber [23], Levine and Orr [60] and Ranicki [84, Introduction].

**2.1. Seifert Surfaces.** It is well-known that every knot $S^n \subset S^{n+2}$ is the boundary of a connected and oriented $(n+1)$-manifold

(2) $$S^n = \partial V^{n+1} \subset V^{n+1} \subset S^{n+2}$$

called a *Seifert surface*. Although there are (infinitely) many possible choices for $V^{n+1}$, a 'Seifert matrix' of linking numbers associated to the embedding of any Seifert surface in $S^{n+2}$ enable one to define and compute valuable isotopy invariants and cobordism invariants of the knot. Indeed, M.Kervaire [41] and J.Levine [56] proved that when $n \geq 2$ complete cobordism information is obtained. We elaborate in section 4 below.

One can explain both the existence of Seifert surfaces and the independence of associated invariants to one's choice of Seifert surface as follows: Let $X$ denote the exterior of a knot $K$, that is to say, the complement in $S^{n+2}$ of an open tubular neighbourhood of $K$. By Alexander duality $H_*(X) \cong H_*(S^1)$ for all knots $K$ so there is a canonical surjection

$$\pi_1(X) \to \pi_1(X)^{\mathrm{ab}} \cong H_1(X) \cong \mathbb{Z} . \tag{3}$$

This surjection is induced by a map

$$\theta : X \to S^1$$

which sends a meridian of the knot to a generator of $\pi_1(S^1) = \mathbb{Z}$ and is unique up to homotopy. If $x \in S^1$ and $\theta$ is chosen judiciously then the inverse image $\theta^{-1}(x)$ (together with a small collar) is a Seifert surface for $K$.

**2.2. The Infinite Cyclic Cover.** Although Seifert surfaces facilitate computation, a knot invariant is perhaps more elegant if its definition does not require arbitrary choices. J.Milnor [**68**] and R.Blanchfield [**7**] considered the algebraic topology of an infinite cyclic cover $\overline{X} \to X$, the pull-back of $\theta : X \to S^1$

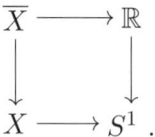

The space $\overline{X}$ enjoys a free action by the group $\mathbb{Z}$ of deck transformations. Poincaré duality of $\overline{X}$, defined $\mathbb{Z}$-equivariantly, yields knot invariants which are equivalent in sensitivity to 'Seifert matrix' invariants. In particular, when $n \geq 2$ one can obtain complete knot cobordism invariants. We shall outline the various approaches to knot cobordism in section 4.

## 3. Boundary Links

One would naturally like to extend the knot theory we have outlined to links; the reader is referred to the recent book of J.Hillman [**36**] for a broad treatment of the algebraic theory.

Certainly one can obtain link invariants by means of a map $\theta : X \to S^1$ where $X$ denotes the link exterior. If $\theta$ is chosen suitably then the preimage of a point $x \in S^1$ is a connected oriented $(n+1)$-manifold which spans the link (compare Tristram [**98**]).

A link admits not only an infinite cyclic cover, but also a free abelian cover. By Alexander duality $H_1(X) \cong \mathbb{Z}^\mu$ so there is a natural surjection

$$\pi_1(X) \to \pi_1(X)^{\mathrm{ab}} \cong H_1(X) \cong \mathbb{Z}^\mu .$$

which leads to further link invariants (e.g. Hillman [**35**] and Sato [**89, 91**]).

Closer parallels to high-dimensional knot theory become possible if one assumes that there is a suitable homomorphism from $\pi_1(X)$ to the (non-abelian) free group $F_\mu$ on $\mu$-generators. Such a homomorphism exists if and only if the components of the link bound *disjoint* oriented $(n+1)$-manifolds.

DEFINITION 1.3. A link $L$ is a *boundary link* if the components bound disjoint connected oriented $(n+1)$-manifolds

$$L = \partial(V_1^{n+1} \sqcup \cdots \sqcup V_\mu^{n+1}) \subset V_1^{n+1} \sqcup \cdots \sqcup V_\mu^{n+1} \subset S^{n+2}.$$

The union $V$ of these $(n+1)$-manifolds is called a *Seifert surface*.

PROPOSITION 1.4 (Gutierrez [33], Smythe [96]). *A link $L$ is a boundary link if and only if there is a group homomorphism*

$$\theta : \pi_1 X \to F_\mu$$

*which sends some choice of meridians $m_1, \cdots, m_\mu$ of the components of $L$ to a distinguished basis $z_1, \cdots, z_\mu$ for $F_\mu$.*

DEFINITION 1.5. A pair $(L, \theta)$ as in theorem 1.4 is called an $F_\mu$-*link*.

In particular, the map (3) above implies that every knot is a boundary link, as we discussed in section 2.1. A knot can also be regarded as an $F_1$-link because (3) is unique.

Meridians $m_i \in \pi_1(X)$ in theorem 1.4 can be defined by choosing a suitable path from a base-point of $X$ to a small meridinal circle around each component of the link. Two surjections $\theta$ and $\theta'$ corresponding to distinct choices of meridians are related by an equation $\theta = \alpha \theta'$ where $\alpha$ is an automorphism of $F_\mu$ which sends each generator $z_i$ to some conjugate $g_i z_i g_i^{-1} \in F_\mu$.

Returning briefly to knot theory, the Thom-Pontrjagin construction implies that every Seifert surface is $\theta^{-1}(x)$ for some $\theta : X \to S^1$ and $x \in S^1$. Given two Seifert surfaces $V = \theta^{-1}(x)$ and $V' = \theta'^{-1}(x)$ one can find a homotopy $h : X \times [0,1] \to S^1$ from $\theta$ to $\theta'$. If $h$ is chosen judiciously, then $h^{-1}(x)$ is a cobordism between $V$ and $V'$ relative the knot.

Boundary link theory is similar: One constructs a Seifert surface from a map not to a circle but to a one-point union of $\mu$ circles

$$X \to S^1 \vee \cdots \vee S^1.$$

The inverse image of $\mu$ regular points chosen on distinct copies of $S^1$ is a $\mu$-component Seifert surface. Conversely, given a Seifert surface the Thom-Pontrjagin construction gives a map $X \to S^1 \vee \cdots \vee S^1$. The choices of surjection $\theta : \pi_1(X) \to F_\mu$ correspond bijectively with the cobordism classes (rel L) of Seifert surfaces.

**3.1. Cobordisms.** Just as 'link' is not the only generalization of 'knot', 'link cobordism' is not the only generalization of 'knot cobordism'.

DEFINITION 1.6. Two boundary links $L^0$ and $L^1$ are *boundary cobordant* if they can be joined in $S^{n+2} \times [0,1]$ by a link cobordism $LI$ whose components bound disjoint oriented $(n+2)$-manifolds. If a boundary link is boundary cobordant to the trivial link, it is called *boundary-slice*. The set of boundary cobordism classes of boundary links is denoted $B(n, \mu)$.

DEFINITION 1.7. Two $F_\mu$-links $(L^0, \theta_0)$ and $(L^1, \theta_1)$ are *cobordant* if they may be joined in $S^{n+2} \times [0, 1]$ by a pair $(LI, \Theta)$ where $LI$ is a link cobordism and
$$\Theta : \pi_1(S^{n+2} \times [0, 1] \setminus LI) \to F_\mu$$
agrees with $\theta$ and $\theta'$ up to inner automorphism. The set of cobordism classes of $F_\mu$-links is denoted $C_n(F_\mu)$.

In the case of knot theory, $\mu = 1$, every knot cobordism extends to a cobordism between Seifert surfaces. In symbols
$$C_n(F_1) = B(n, 1) = C(n, 1) .$$

Returning to links, there are canonical forgetful maps
$$C_n(F_\mu) \to B(n, \mu) \to C(n, \mu)$$
but they are not in general bijective. T.Cochran and K.Orr proved [**15, 16**] that if $n$ is odd and $\mu \geq 2$ then $B(n, \mu) \to C(n, \mu)$ is not surjective. It is an open problem to determine whether this map is injective.

The map $C_n(F_\mu) \to B(n, \mu)$ is easier to understand. As we discussed in section 3, a boundary link has a splitting $\theta : \pi_1(X) \to F_\mu$ defined uniquely up to composition by generator conjugating automorphisms of the free group. If $A_\mu$ denotes the group of such automorphisms (modulo inner automorphisms) then $B(n, \mu)$ is the set of $A_\mu$-orbits
$$B(n, \mu) = C_n(F_\mu)/A_\mu .$$
Note that $A_\mu$ is a non-trivial group for all $\mu \geq 3$.

Ko proved further [**45**, Theorem 2.7] that there is a bijective correspondence between cobordism classes of $F_\mu$-links and cobordism classes of pairs
$$(V^{n+1} \subset S^{n+2},\ L = \partial V) = \text{(Seifert Surface, Boundary Link)}.$$
He also gave a geometric interpretation of the action of $A_\mu$ in terms of Seifert surfaces.

The present work uses Seifert surface methods to define invariants which distinguish cobordism classes of $F_\mu$-links.

**3.2. Addition of $F_\mu$-Links.** It is well-known that one can add two knots by performing an ambient connected sum. Picturing the two knots embedded, but separated, in a single ambient space $S^{n+2}$, one chooses an arc in $S^{n+2}$ joining the two knots. One cuts out a small disk $D^n$ from each knot $S^n$ and attaches the knots by a narrow tube $S^{n-1} \times [0, 1]$ which surrounds the arc. The isotopy class of this sum is independent of the choice of connecting arc and the set of isotopy classes of knots $S^n \subset S^{n+2}$ becomes a commutative semigroup.

Knot cobordism respects addition of knots; in fact $C(n, 1)$ is an abelian group. The inverse of a knot $S^n \subset S^{n+2}$ is obtained by reversing the orientations both of $S^n$ and of the ambient space $S^{n+2}$.

Addition of links is more delicate. Given links $L^0$ and $L^1$ one can join each component of $L^0$ to the corresponding component of $L^1$ by an arc, but the isotopy class and the cobordism class of the sum usually depend on the choice of arcs. In general, neither $C(n,\mu)$ nor $B(n,\mu)$ have an obvious group structure.

On the other hand, there is a well-defined notion of addition for two $F_\mu$-links $(L^0, \theta_0)$ and $(L^1, \theta_1)$ if one assumes that $\theta_0$ and $\theta_1$ are isomorphisms. Denoting $z_1, \cdots, z_\mu$ a distinguished basis of $F_\mu$, one can use preferred meridians $\theta_0^{-1}(z_j)$ and $\theta_1^{-1}(z_j)$ to define a connecting arc between the $j$th component of $L^0$ and the $j$th component of $L^1$ (cf Le Dimet [**52**]).

Equivalently, one may choose simply-connected Seifert surfaces corresponding to $\theta_0$ and $\theta_1$ and then join the components of $L^0$ to components of $L^1$ along arcs which avoid both Seifert surfaces.

In the classical dimension, $C_1(F_\mu)$ does not have an obvious group structure if $\mu \geq 2$. However, if $n \geq 2$ then every $F_\mu$-link $(L, \theta)$ is cobordant to some $(L', \theta')$ such that $\theta'$ is an isomorphism. Moreover, cobordism respects addition of $F_\mu$-links so $C_n(F_\mu)$ is an abelian group for all $n \geq 2$. For further details see Ko [**45**, Prop 2.11] or Mio [**71**, p260].

**3.3. Split $F_\mu$-links.** Given $\mu$ knots there is a canonical way to make a link; one puts all the knots in a single ambient space $S^{n+2}$, far apart from each other so that the components lie in disjoint disks $D^{n+2}$. A link which can be constructed in this way is called a *split* link. Choosing a Seifert surface for each knot, one obtains a Seifert surface for the split link. It follows that a split link is, canonically, an $F_\mu$-link $(L, \theta)$ where $\theta: \pi_1(X) \to F_\mu$ is the free product of the canonical homomorphisms (3) associated to the component knots. We thus have a canonical map

$$\text{(4)} \qquad \prod_{i=1}^{\mu} C_n(F_1) \to C_n(F_\mu)$$

which is a group homomorphism if $n \geq 2$.

Conversely, there are $\mu$ forgetful maps $C_n(F_\mu) \to C_n(F_1)$, which, together, split (4). If $n \geq 2$ there is therefore a decomposition

$$C_n(F_\mu) \cong \left( \bigoplus_{i=1}^{\mu} C_n(F_1) \right) \oplus \widetilde{C}_n(F_\mu) .$$

We shall return to the $F_\mu$-link cobordism group in section 5, for we must first describe the reduction to algebra of high-dimensional knot cobordism $C_n(F_1)$.

## 4. Surgery Obstructions

Given a high-dimensional knot, one can attempt to construct a cobordism to the trivial knot by performing surgeries. It turns out that, in even dimensions, such a feat is always possible; every knot is null-cobordant.

In odd dimensions, obstructions exist but the attempt leads to an identification of $C_{2q-1}(F_1)$ with a 'surgery obstruction group'.

### 4.1. Surgery on a Seifert Surface.

THEOREM 1.8 (Kervaire 1965). *Every even-dimensional knot* $K \cong S^{2q} \subset S^{2q+2}$ *is null-cobordant:*

$$C_{2q}(F_1) = 0 \text{ for all } q \geq 1.$$

SKETCH PROOF. Starting with any Seifert surface

$$V^{2q+1} \subset S^{2q+2}, \quad \partial V = K$$

one performs surgery on $V^{2q+1}$ killing homology classes of degree at most $q$, until one has turned $V$ into a disk $D^{2q+1}$. One must check that the surgery operations can be performed ambiently, not in $S^{2q+2}$, but in a disk $D^{2q+3}$ whose boundary is $S^{2q+2}$.

The effect of each surgery operation is to cut out a copy of $S^i \times D^{2q-i}$ in the interior of $V^{2q}$ and graft a copy of $D^{i+1} \times S^{2q-i-1}$ along the boundary $S^i \times S^{2q-i-1}$. At core, such surgeries are possible because the embedded spheres $S^i$ bound disks $D^{i+1} \subset D^{2q+3}$ which are disjoint and intersect $V$ only at $S^i$. □

The same method shows that every even-dimensional boundary link is boundary-slice, a result which was proved by Cappell and Shaneson [12] using homology surgery (see sections 4.4 and 5.1). In symbols,

$$C_{2q}(F_\mu) = B(2q, \mu) = 0 \text{ for all } \mu \geq 1 \text{ and all } q \geq 1.$$

By contrast not all odd-dimensional knots $K = S^{2q-1} \subset S^{2q+1}$ are slice. In fact the odd-dimensional knot cobordism groups are not even finitely generated. If $V^{2q}$ is a Seifert surface for $K$ one can perform ambient surgery to kill all homology in degree strictly less than $q$, but $q$-spheres, if they are linked in $S^{2q+1}$, do not bound disjoint disks $D^{q+1}$ in $D^{2q+1}$. The Seifert form measures this obstruction to surgery in dimension $q$ and was used by J.Levine [56] to obtain an algebraic description of odd-dimensional knot cobordism groups (theorem 1.10 below).

### 4.2. The Seifert Form.
By Alexander duality there is a non-singular pairing which measures linking between $V^{2q}$ and $S^{2q+1}\backslash V^{2q}$:

$$H_q(V) \times H_q(S^{2q+1}\backslash V) \to \mathbb{Z}$$
$$(x, x') \mapsto \text{Lk}(x, x') .$$

We may assume, after preliminary surgeries below the middle dimension, that $H_q(V^{2q})$ is a free abelian group. Let $x_1, \cdots, x_m$ be a basis for $H_q(V)$ and let $i_+, i_- : H_q(V^{2q}) \to H_q(S^{2q+1}\backslash V^{2q})$ be small translations in the positive and negative normal directions to $V^{2q}$. These directions are determined by the orientation of $V$.

## 4. SURGERY OBSTRUCTIONS

DEFINITION 1.9. The *Seifert matrix* $S$ associated to the Seifert surface $V$ with respect to $x_1, \cdots, x_m$ is the matrix
$$S_{ij} = \mathrm{Lk}(i_+ x_i, x_j) = \mathrm{Lk}(x_i, i_- x_j) .$$

Addition of knots $K + K'$ by connected sum induces a block sum $\begin{pmatrix} S & 0 \\ 0 & S' \end{pmatrix}$ of Seifert matrices. Although the Seifert matrix $S$ is not symmetric or skew-symmetric, $S + (-1)^q S^t$ is certainly $(-1)^q$-symmetric (where $S^t$ denotes the transpose of $S$). It is not difficult to see that
$$S_{ij} + (-1)^q S_{ji} = \mathrm{Lk}((i_+ - i_-)x_i, x_j)$$
which is the intersection pairing on $H_q(V^{2q})$. By Poincaré duality this intersection matrix $S + (-1)^q S^t$ is non-singular[4].

Kervaire proved [**41**, Théorème II.3] that every matrix $S$ such that $S + (-1)^q S^t$ is invertible is the Seifert matrix of some Seifert surface of some knot $S^{2q-1} \subset S^{2q+1}$ (when $q \neq 2$). Note that if one working in the category of smooth manifolds the differentiable structure on $S^{2q-1}$ may be exotic (see footnote 2 on page 2). To outline Kervaire's argument, one can first construct a stably parallelizable manifold $V^{2q}$ with intersection pairing $S + (-1)^q S^t$ by attaching $q$-handles $D^q \times D^q$ to the boundary of a zero-handle $D^{2q}$. Since $S + (-1)^q S^t$ is invertible, the boundary of $V^{2q}$ is a homotopy sphere, and hence is homeomorphic to a sphere. One can complete the construction by adjusting an embedding $V^{2q} \hookrightarrow S^{2q+1}$ until the handles are linked in the manner dictated by $S$.

If a $(2q-1)$-dimensional knot $K$ is slice then one can find a basis for $H_q(V)$ half of which is completely unlinked in $S^{2q+1}$. In other words, the Seifert matrix has the appearance

(5)
$$\begin{pmatrix} 0 & * \\ * & * \end{pmatrix}$$

where each $*$ denotes some square matrix with half the rank of $H_q(V)$. Such a Seifert matrix is called *metabolic*. It follows that there is a group homomorphism from $C_{2q-1}(F_1)$ to the Witt group $G^{(-1)^q,1}(\mathbb{Z})$ of Seifert matrices modulo metabolic matrices (see section 4 of chapter 3 for a more precise definition).

If $q \geq 2$ the converse is true: Any knot $S^{2q-1} \subset S^{2q+1}$ which has a metabolic Seifert form is a slice knot.

THEOREM 1.10 (Levine 1969). *If* $q \geq 3$ *then* $C_{2q-1}(F_1) \cong G^{(-1)^q,1}(\mathbb{Z})$.

In the case $q = 2$, the cobordism group $C_3(F_1)$ is isomorphic to an index two subgroup of $G^{1,1}(\mathbb{Z})$. If $q = 1$ there is a surjection $C_1(F_1) \twoheadrightarrow G^{-1,1}(\mathbb{Z})$ but A.Casson and C.Gordon defined knot invariants [**13, 14**] which show that the kernel is non-trivial (see also Gilmer [**31, 32**], Kirk and Livingston [**43**] and Letsche [**53**]). T.Cochran, K.Orr and P.Teichner [**17**] have defined an infinite tower of obstructions to slicing a knot and have recovered

---
[4]The exact sequence for the pair $(V, \partial V)$ implies that $H_q(V, \partial V) \cong H_q(V)$.

all the earlier cobordism invariants in the early layers. At the time of writing, it remains an open problem to complete the cobordism classification of classical knots.

**4.3. The Blanchfield Form.** As we indicated in section 2.2 the infinite cyclic cover $\overline{X}^{2q+1}$ of an odd-dimensional knot exterior $X^{2q+1}$ can be used to define an algebraic obstruction to slicing a knot without reference to a choice of Seifert surface.

Since $\mathbb{Z}$ acts on $\overline{X}$, the homology $H_i(\overline{X})$ is a module over the group ring $\mathbb{Z}[\mathbb{Z}] = \mathbb{Z}[z, z^{-1}]$. We may assume, if necessary by taking a quotient by the $\mathbb{Z}$-torsion subgroup, that $H_q(\overline{X})$ is $\mathbb{Z}$-torsion-free and has a one-dimensional presentation

$$(6) \qquad 0 \to \mathbb{Z}[z, z^{-1}]^m \xrightarrow{\sigma} \mathbb{Z}[z, z^{-1}]^m \to H_q\overline{X} \to 0.$$

The determinant $p = \det(\sigma) \in \mathbb{Z}[z, z^{-1}]$ of such a presentation is called the *Alexander polynomial* of the knot and is well-defined up to multiplication by units $\pm z^i$. Since $H_*(X; \mathbb{Z}) \cong H_*(S^1; \mathbb{Z})$, one finds that $\sigma$ becomes an invertible matrix over $\mathbb{Z}$ if one sets $z = 1$. In other words $p(1) = \pm 1$.

By Poincaré duality[5] there is a non-singular 'Blanchfield linking form':

$$(7) \qquad \beta : H_q(\overline{X}) \times H_q(\overline{X}) \to \frac{P^{-1}\mathbb{Z}[z, z^{-1}]}{\mathbb{Z}[z, z^{-1}]}$$

where $P = \{p \in \mathbb{Z}[z, z^{-1}] \mid p(1) = \pm 1\}$ is the set of Alexander polynomials. The ring $\mathbb{Z}[z, z^{-1}]$ has an involution $\overline{z} = z^{-1}$ with respect to which the Blanchfield form is $(-1)^{q+1}$-hermitian:

$$\beta(x, x') = (-1)^{q+1}\overline{\beta(x', x)} \text{ for all } x, x' \in H_q(\overline{X}).$$

C.Kearton showed [39] that a Seifert form for a knot is metabolic if and only if the Blanchfield form of the knot is metabolic. He also proved [38, Theorem 8.3] that if $q \neq 2$ then every $\mathbb{Z}[z, z^{-1}]$-module with presentation (6) and a linking pairing (7) is the Blanchfield form of some knot $S^{2q-1} \subset S^{2q+1}$. Consequently,

THEOREM 1.11 (Kearton, Levine 1975). *If $q \geq 3$ then $C_{2q-1}(F_1)$ is isomorphic to the Witt group of $(-1)^{q+1}$-hermitian Blanchfield forms.*

**4.4. Surgery and Homology Surgery.** A third approach to knot cobordism was pioneered by S.Cappell and J.Shaneson [11]. We shall give a simplified outline of their method, but we must first review a little of the original surgery theory of Browder, Novikov, Sullivan, Wall and many others.

One of the central questions in surgery theory is the following

QUESTION 1.12. *Given a homotopy equivalence $f : N^n \to M^n$ between two manifolds, is $f$ is homotopic to a homeomorphism?*

---

[5]Note that $H_q(\overline{X}, \partial\overline{X}) \cong H_q(\overline{X})$.

One can attack the problem in two stages. Firstly one seeks a (normal) bordism $F: W^{n+1} \to M \times [0,1]$ such that $\partial W = N \sqcup N'$ and such that $F$ restricts to $f: N \to M \times \{0\}$ and to a homeomorphism $N' \to M \times \{1\}$. Secondly, if such a bordism $F$ exists, one tries to change it, by surgeries, to make it a (simple) homotopy equivalence. In other words, one asks whether $F$ is bordant, relative the boundary, to a homotopy equivalence. If these two steps can both be achieved then the $s$-cobordism theorem implies that $f$ is homotopic to a homeomorphism.

Without dwelling on the potential obstruction to the first stage, let us note that there is an obstruction $\sigma(F)$ to the second stage of the program which takes values in a group $L_{n+1}(\mathbb{Z}[\pi_1(M)])$ introduced by C.T.C.Wall [101]. The map $f$ is homotopic to a homeomorphism if and only if one can find a bordism $F$ such that $\sigma(F) = 0 \in L_{n+1}(\mathbb{Z}[\pi_1(M)])$.

In A.Ranicki's algebraic surgery theory [81, 82, 83], $L_{n+1}(\mathbb{Z}[\pi])$ is the cobordism group of $(n+1)$-dimensional free $\mathbb{Z}[\pi]$-module chain complexes with (quadratic) Poincaré duality structure. Even-dimensional $L$-groups are isomorphic to Witt groups of quadratic forms whereas the odd-dimensional groups can be expressed in terms of 'formations'.

Returning to knot theory, the question whether a knot is null-cobordant has a formulation analogous to 1.12. The exterior $X^{n+2}$ of a knot $K$ is homotopic to the trivial knot exterior $X_0^{n+2}$ if and only if $K$ is trivial (Levine [54], Wall [101, Theorem 16.4]). On the other hand, *any* knot exterior $X^{n+2}$ admits a homology equivalence[6] $\theta: X \to X_0$ which restricts to a homotopy equivalence $\partial X \to \partial X_0$.

Conversely, suppose we are given a manifold $X^{n+2}$ with the properties that $\partial X = S^1 \times S^n$ and $\pi_1(X)$ is normally generated by $\pi_1(\partial X) = \mathbb{Z}$. If $H_*(X) = H_*(X_0)$ then one can regard $X^{n+2}$ as a knot exterior by gluing a solid torus $D^2 \times S^n$ along the boundary to form a sphere $S^{n+2}$, in which the core of the torus $\{0\} \times S^n$ is a knot. The exterior of a knot cobordism has a similar characterization so the question whether a knot $K$ is null-cobordant translates as follows:

QUESTION 1.13. *Given a homology equivalence $\theta: X \to X_0$, is $\theta$ homology bordant to a homotopy equivalence?*

A homology bordism between $\theta: X \to X_0$ and $\theta': X' \to X_0$ is by definition a homology equivalence $\Theta: W \to X_0$ such that $\partial W = X \sqcup X'$, and such that $\Theta|_X = \theta$ and $\Theta|_{X'} = \theta'$.

One can tackle the problem in two stages. Firstly one seeks any (normal) bordism $\Theta: W \to X_0$ between $\theta: X \to X_0$ and a homotopy equivalence $\theta': X' \to X_0$. Secondly one tries to change the bordism $\Theta$, by surgeries, to make it a homology equivalence. There are obstructions at each stage; the first takes values in

$$L_{n+2}(\mathbb{Z}[\pi_1(X_0)]) = L_{n+2}(\mathbb{Z}[z, z^{-1}])$$

---

[6]To be more precise, one begins with a degree one normal map whose restriction to boundaries is a homotopy equivalence $\partial X \simeq \partial X_0 \simeq S^n \times S^1$.

while the second lies in a group

$$\Gamma_{n+3}(\epsilon : \mathbb{Z}[z,z^{-1}] \to \mathbb{Z})$$

defined by Cappell and Shaneson. Here, $\epsilon$ is the augmentation map which evaluates polynomials at $z = 1$.

In the framework of Ranicki [**83**, §2.4], $\Gamma_n(\mathbb{Z}[z,z^{-1}] \to \mathbb{Z})$ is the cobordism group of free $n$-dimensional $\mathbb{Z}[z,z^{-1}]$-module chain complexes with a quadratic structure such that the induced structure over $\mathbb{Z}$ is Poincaré.

The knot $K$ is null-cobordant if and only if one can find a bordism $\Theta$ between $\theta$ and a homotopy equivalence such that the surgery obstruction $\sigma(\Theta) \in \Gamma_{n+3}(\epsilon : \mathbb{Z}[z,z^{-1}] \to \mathbb{Z})$ vanishes. These obstructions, together with an appropriate realization theorem, imply that the various cobordism groups fit together in an exact sequence:

$$(8) \quad \cdots \to L_{n+3}(\mathbb{Z}[z,z^{-1}]) \to \Gamma_{n+3}\left(\mathbb{Z}[z,z^{-1}] \xrightarrow{\epsilon} \mathbb{Z}\right) \to C_n(F_1) \to L_{n+2}(\mathbb{Z}[z,z^{-1}]) \to \cdots$$

Identifying $L_*(\mathbb{Z}[z,z^{-1}]) \cong \Gamma_*(\mathbb{Z}[z,z^{-1}] \to \mathbb{Z}[z,z^{-1}])$ one obtains:

THEOREM 1.14 (Cappell and Shaneson 1974). *If $n \neq 1$ or $3$ then $C_n(F_1)$ is isomorphic to a relative $\Gamma$-group:*

$$C_n(F_1) \cong \Gamma_{n+3}\begin{pmatrix} \mathbb{Z}[z,z^{-1}] \xrightarrow{\mathrm{id}} \mathbb{Z}[z,z^{-1}] \\ \mathrm{id} \downarrow \qquad\qquad \downarrow \epsilon \\ \mathbb{Z}[z,z^{-1}] \xrightarrow{\epsilon} \mathbb{Z} \end{pmatrix}.$$

Cappell and Shaneson also gave a geometric interpretation of the 4-periodicity $C_n(F_1) \cong C_{n+4}(F_1)$, $n \geq 4$ of high-dimensional knot cobordism groups [**11**, Theorem 13.3].

The relationship between theorems 1.14 and 1.11 was explained by Pardon [**76**, **77**], Ranicki [**83**, §7.9] and Smith [**95**]. Recalling that $P = \epsilon^{-1}\{\pm 1\} \subset \mathbb{Z}[z,z^{-1}]$ is the set of Alexander polynomials, the Witt group of Blanchfield forms in theorem 1.11 is isomorphic to a relative $L$-group:

$$C_n(F_1) \cong L_{n+3}(\mathbb{Z}[z,z^{-1}], P) \text{ for all } n \neq 1 \text{ or } 3$$

and fits into a localization exact sequence

$$(9) \quad \cdots \to L_{n+3}(\mathbb{Z}[z,z^{-1}]) \to L_{n+3}(P^{-1}\mathbb{Z}[z,z^{-1}]) \to C_n(F_1) \to L_{n+2}(\mathbb{Z}[z,z^{-1}]) \to \cdots$$

This sequence is isomorphic to (8) for there is a natural isomorphism

$$(10) \quad L_*(P^{-1}\mathbb{Z}[z,z^{-1}]) \cong \Gamma_*(\mathbb{Z}[z,z^{-1}] \to \mathbb{Z}).$$

A systematic approach to the algebraic $K$- and $L$-theory which pertains to high-dimensional knots and other manifold embeddings can be found in Ranicki [84][7].

## 5. Algebraic $F_\mu$-link Cobordism

**5.1. Homology Surgery.** The first identification of $F_\mu$-link cobordism groups with surgery obstruction groups was obtained by S.Cappell and J.Shaneson [12] using homology surgery methods.

The exterior $X$ of a boundary link $L$ is homotopy equivalent to the trivial link exterior $X_0$ if and only if $L$ is isotopic to the trivial link (Gutierrez [33]). On the other hand, all link exteriors are homology equivalent, by Alexander duality, and a boundary link exterior admits a homology equivalence $\theta : X \to X_0$ which restricts to a homotopy equivalence on each boundary component $S^1 \times S^n$ (cf theorem 1.4).

Conversely, suppose we are given a manifold $X^{n+2}$ with boundary $\partial X = \bigsqcup_\mu S^1 \times S^n$ such that $\pi_1(X)$ is normally generated by the fundamental groups of the boundary components. If $\theta : X \to X_0$ is a homology equivalence which is a homotopy equivalence on boundary components, one can recover an $F_\mu$-link by gluing $\mu$ copies of $D^2 \times S^n$ onto the boundary of $X$.

The question whether an $F_\mu$-link is cobordant to the trivial link translates into the question whether $\theta : X \to X_0$ is homology bordant to a homotopy equivalence.

THEOREM 1.15 (Cappell and Shaneson 1980). *If $n \neq 1,3$ and $\mu \geq 1$ there is an isomorphism*

$$C_n(F_\mu) \cong \Gamma_{n+3} \begin{pmatrix} \mathbb{Z}[F_\mu] \xrightarrow{\text{id}} \mathbb{Z}[F_\mu] \\ \text{id} \downarrow \qquad \downarrow \epsilon \\ \mathbb{Z}[F_\mu] \xrightarrow{\epsilon} \mathbb{Z} \end{pmatrix}$$

*where the augmentation $\epsilon : \mathbb{Z}[F_\mu] \to \mathbb{Z}$ sends every element of the free group to 1.*

In other words, $C_n(F_\mu)$ fits into a long exact sequence

(11)
$$\cdots \to L_{n+3}(\mathbb{Z}[F_\mu]) \to \Gamma_{n+3}\left(\mathbb{Z}[F_\mu] \xrightarrow{\epsilon} \mathbb{Z}\right)$$
$$\to C_n(F_\mu) \to L_{n+2}(\mathbb{Z}[\mu]) \to \cdots$$

There is a two-stage obstruction to the existence of a homology bordism between $\theta$ and a homotopy equivalence just as in knot theory (section 4.4). The first is the Wall obstruction

$$\sigma(\theta) \in L_{n+2}(\mathbb{Z}[\pi_1(X_0)]) = L_{n+2}(\mathbb{Z}[F_\mu]) = \Gamma_{n+2}(\mathbb{Z}[F_\mu] \to \mathbb{Z}[F_\mu])$$

---

[7]To translate notation, the Witt group $G^{(-1)^q,1}(A)$ of Seifert forms is denoted $\text{LIso}^{2q}(A)$ in [84].

to the existence of any (normal) bordism $\Theta$ while the second is the homology surgery obstruction

$$\sigma(\Theta) \in \Gamma_{n+3}(\mathbb{Z}[F_\mu] \xrightarrow{\epsilon} \mathbb{Z}).$$

**5.2. The $\mu$-component Seifert Form.** K.H.Ko [45] and W.Mio [71] described the odd-dimensional groups $C_{2q-1}(F_\mu)$ in terms of $\mu$-component Seifert forms. Their results will be our starting point. To illustrate the definitions let us consider boundary links of two components for the general case is very similar (see section 4 in chapter 3).

The homology of a two component Seifert surface is a direct sum $H_q(V) = H_q(V_1) \oplus H_q(V_2)$. If one chooses a basis for each summand $H_q(V_i)$ then one obtains a Seifert matrix

$$S = \left(\begin{array}{c|c} S^{11} & S^{12} \\ \hline S^{21} & S^{22} \end{array}\right)$$

where the entries of $S^{ij}$ are linking numbers between elements of $H_q(V_i)$ and $H_q(V_j)$ in the sphere $S^{2q+1}$. The intersection pairing of the Seifert surface is given, as before, by the $(-1)^q$-symmetric non-singular matrix

$$T = \left(\begin{array}{c|c} T_1 & 0 \\ \hline 0 & T_2 \end{array}\right) = S + (-1)^q S^t .$$

In knot theory, two Seifert matrices $S$ and $S'$ are congruent if and only if $S = PS'P^t$ for some invertible matrix $P$. In boundary link theory, any change of basis must preserve the component structure. Two 2-component Seifert matrices $S$ and $S'$ are congruent if and only if $S = PS'P^t$ where $P = \left(\begin{array}{c|c} P_1 & 0 \\ \hline 0 & P_2 \end{array}\right)$ and $P_1$ and $P_2$ are invertible.

The definition of metabolic must be refined accordingly. A Seifert matrix is metabolic if and only if it is congruent to a matrix of the form

$$PSP' = \left(\begin{array}{cc|cc} 0 & * & 0 & * \\ * & * & * & * \\ \hline 0 & * & 0 & * \\ * & * & * & * \end{array}\right)$$

where the four matrices in the $ij$th block each have half as many columns and half as many rows as $S^{ij}$. The Witt group of $\mu$-component Seifert matrices will be denoted $G^{(-1)^q,\mu}(\mathbb{Z})$.

A handlebody construction shows that if $q \geq 3$ then every $\mu$-component Seifert matrix is realised by some $(2q-1)$-dimensional boundary link. Levine's theorem 1.10 generalizes as follows:

THEOREM 1.16 (Ko, Mio 1987). *For all $\mu \geq 1$ and for all $q \geq 3$,*

$$C_{2q-1}(F_\mu) \cong G^{(-1)^q,\mu}(\mathbb{Z}).$$

Ko also studied the cases $q = 2$ and $q = 1$ showing that $C_3(F_\mu)$ is isomorphic to an index $2^\mu$ subgroup of $G^{1,\mu}(\mathbb{Z})$ while $C_1(F_\mu)$ maps onto

### 5.3. The Blanchfield-Duval Form.

J.Duval [22] generalized to $F_\mu$-links the Blanchfield forms of odd-dimensional knot theory. In place of the infinite cyclic cover of a knot exterior one studies the covering $\overline{X}$ of a boundary link exterior induced by $\theta : \pi_1(X) \to F_\mu$. Unlike knot theory, the middle homology $H_q(\overline{X})$ is such that $\mathbb{Q} \otimes_{\mathbb{Z}} H_q(\overline{X})$ is infinite-dimensional unless it is trivial. However, taking a quotient by the torsion subgroup is necessary we may assume that $H_q(\overline{X})$ is $\mathbb{Z}$-torsion-free and has a presentation

$$0 \to \mathbb{Z}[F_\mu]^m \xrightarrow{\sigma} \mathbb{Z}[F_\mu]^m \to H_q(\overline{X}) \to 0$$

where $\sigma$ is a square matrix which becomes invertible under the augmentation $\epsilon : \mathbb{Z}[F_\mu] \to \mathbb{Z}$ (cf (6) above and Sato [90]). Let $\Sigma$ denote the set of such $\epsilon$-invertible matrices.

Duval showed that the $F_\mu$-equivariant Poincaré duality of $\overline{X}$ gives rise to a linking form (compare Farber [26])

$$H_q(\overline{X}) \times H_q(\overline{X}) \to \frac{\Sigma^{-1}\mathbb{Z}[F_\mu]}{\mathbb{Z}[F_\mu]}$$

where $\Sigma^{-1}\mathbb{Z}[F_\mu]$ is the Cohn localization, defined by formally adjoining inverses to all the matrices in $\Sigma$. M.Farber and P.Vogel [27] subsequently identified $\Sigma^{-1}\mathbb{Z}[F_\mu]$ with a ring of rational power series in $\mu$ non-commuting indeterminates.

THEOREM 1.17 (Duval 1986). *If $q \geq 3$ then $C_{2q-1}(F_\mu)$ is isomorphic to the Witt group of Blanchfield-Duval forms.*

It follows that $C_{2q-1}(F_\mu)$ is isomorphic to $L_{2q+2}(\mathbb{Z}[F_\mu], \Sigma)$, the relative $L$-group in the localization exact sequence of Vogel [99, 100]:

$$\cdots \to L_{n+3}(\mathbb{Z}[F_\mu]) \to L_{n+3}(\Sigma^{-1}\mathbb{Z}[F_\mu]) \to C_n(F_\mu) \to L_{n+2}(\mathbb{Z}[F_\mu]) \to \cdots$$

This $L$-theory exact sequence is isomorphic to the exact sequence (11) above (see Duval [22, pp633-634]).

By algebraic means M.Farber constructed [24, 25] self-dual finite rank lattices inside a link module $H_q(\overline{X})$ to mimic the geometric relationship between the Seifert surfaces of a boundary link and the free cover of the link exterior.

CHAPTER 2

# Main Results

The title of the present volume echoes that of a paper by J.Levine [55] which gave an algorithm to decide whether or not two $(2q-1)$-dimensional knots are cobordant, assuming $q>1$; we shall define an algorithm to decide whether or not two $(2q-1)$-dimensional $F_\mu$-links are cobordant, again for $q>1$.

It is a theorem of Kervaire that all the even-dimensional groups are trivial (see chapter 1, section 4.1). In odd dimensions Levine used his knot invariants to prove:

THEOREM 2.1 (Levine 1969). *In every odd dimension $2q-1>1$, the knot cobordism group is isomorphic to a countable direct sum of cyclic groups of orders 2, 4 and $\infty$:*

$$C_{2q-1}(F_1) \cong \mathbb{Z}^{\oplus\infty} \oplus \left(\frac{\mathbb{Z}}{2\mathbb{Z}}\right)^{\oplus\infty} \oplus \left(\frac{\mathbb{Z}}{4\mathbb{Z}}\right)^{\oplus\infty}.$$

Note that any two decompositions of a group into direct sums of cyclic groups of infinite and prime power order are isomorphic (e.g. Fuchs [30, Theorem 17.4]).

The following theorem is a corollary of our main result, theorem B:

THEOREM A. *If $\mu \geq 2$ and $2q-1>1$ then the $F_\mu$-link cobordism group is isomorphic to a countable direct sum of cyclic groups of orders 2, 4, 8 and $\infty$:*

$$C_{2q-1}(F_\mu) \cong \mathbb{Z}^{\oplus\infty} \oplus \left(\frac{\mathbb{Z}}{2\mathbb{Z}}\right)^{\oplus\infty} \oplus \left(\frac{\mathbb{Z}}{4\mathbb{Z}}\right)^{\oplus\infty} \oplus \left(\frac{\mathbb{Z}}{8\mathbb{Z}}\right)^{\oplus\infty}.$$

Levine utilized work of J.Milnor [69] 'On isometries of inner product spaces' to obtain Witt invariants of (1-component) Seifert forms. We shall employ more general non-commutative methods formulated by Quebbemann, Scharlau and Schulte [80] to define Witt-invariants of $\mu$-component Seifert forms. Seifert form invariants correspond to cobordism invariants of knots and $F_\mu$-links by theorems 1.10 and 1.16 above.

One could define $F_\mu$-link cobordism invariants using Blanchfield-Duval forms (theorem 1.17) in place of Seifert forms. The advantage of the Seifert form approach pursued here for computational purposes is that the middle homology of a Seifert surface is finitely generated as a $\mathbb{Z}$-module whereas the middle homology of a free cover of a boundary link complement is not (unless it is zero).

18                                2. MAIN RESULTS

Although the algebraic techniques we use are a direct generalization of Milnor's, our $F_\mu$-link invariants do not directly generalize Levine's knot cobordism invariants; our approach is closer to that of Kervaire [42] (see also Kervaire and Weber [40, p107-111] and Ranicki [84, Chapter 42]).

We shall pay particular attention to signature invariants, which detect the torsion-free part of the $F_\mu$-link cobordism group. In knot theory, there is a signature for each algebraic integer on the line $\{\frac{1}{2} + ai \mid a \in \mathbb{R}\}$, or, equivalently, one for each point on the unit circle which is a root of an Alexander polynomial. We shall define an $F_\mu$-link signature for each 'algebraic integer' on a disjoint union of real algebraic varieties.

## 1. Signatures

Before stating results about $F_\mu$-link signatures, let us first describe the corresponding knot theory.

Recall from section 4.2 of chapter 1 that a Seifert matrix $S$ of linking numbers can be associated to each choice of Seifert surface $V^{2q}$ for a knot $S^{2q-1} \subset S^{2q+1}$. If $M = H_q(V)$ and $\epsilon = (-1)^q$ then $S$ represents a bilinear 'Seifert form' $\lambda : M \to M^*$ such that $\lambda + \epsilon \lambda^*$ is invertible. The Witt group of these Seifert forms with integer entries is denoted $G^{\epsilon,1}(\mathbb{Z})$ and Levine's theorem 1.10 states that $C_{2q-1}(F_1) \cong G^{\epsilon,1}(\mathbb{Z})$ when $q \geq 3$.

**1.1. Complex Coefficients.** A complete set of *torsion-free* invariants of $G^{\epsilon,1}(\mathbb{Z})$ can be obtained by replacing the coefficient ring $\mathbb{Z}$ by $\mathbb{C}^-$. The superscript indicates that we work in the hermitian setting, the involution on $\mathbb{C}$ being complex conjugation. 'Tensoring by $\mathbb{C}^-$' defines a canonical map $G^{\epsilon,1}(\mathbb{Z}) \to G^{\epsilon,1}(\mathbb{C}^-)$.

KNOT THEORY EXAMPLE 2.2. The kernel of the canonical map

(12) $$G^{\epsilon,1}(\mathbb{Z}) \to G^{\epsilon,1}(\mathbb{C}^-)$$

is 4-torsion. Moreover, $G^{\epsilon,1}(\mathbb{C}^-)$ is a free abelian group

$$G^{\epsilon,1}(\mathbb{C}^-) \cong \mathbb{Z}^{\oplus \infty};$$

the points on the line $\{\frac{1}{2} + bi \mid b \in \mathbb{R}\}$ correspond to a basis for $G^{\epsilon,1}(\mathbb{C}^-)$. The image of $G^{\epsilon,1}(\mathbb{Z})$ is contained in the subgroup with one basis element for each algebraic integer $\frac{1}{2} + bi$. The composition of (12) with a projection of $G^{\epsilon,1}(\mathbb{C}^-)$ onto the subgroup generated by a basis element $\omega \in \{\frac{1}{2} + bi \mid b \in \mathbb{R}\}$ is called a signature

$$\sigma_\omega : G^{\epsilon,1}(\mathbb{Z}) \to \mathbb{Z} \, .$$

Note that $\sigma_\omega = \sigma_{\overline{\omega}}$.

We turn to signatures of $F_\mu$-links. Recall that a $\mu$-component Seifert form is a homomorphism $\lambda : M \to M^*$ together with a decomposition $M \cong \pi_1 M \oplus \cdots \oplus \pi_\mu M$. The component $\pi_i M$ denotes the middle homology of the $i$th component of a Seifert surface. Recall further that $G^{\epsilon,\mu}(\mathbb{Z})$ denotes the Witt group of $\mu$-component Seifert forms. Theorem 1.16 states that $C_{2q-1}(F_\mu) \cong G^{\epsilon,\mu}(\mathbb{Z})$ for all $q \geq 3$.

PROPOSITION 2.3. *Let $\mu \geq 1$. The kernel of the canonical map*

(13) $$G^{\epsilon,\mu}(\mathbb{Z}) \to G^{\epsilon,\mu}(\mathbb{C}^-)$$

*is 8-torsion. Moreover, $G^{\epsilon,\mu}(\mathbb{C}^-)$ is a free abelian group.*

**1.2. Varieties of Signatures.** Although the cardinality of $G^{\epsilon,\mu}(\mathbb{C}^-)$ does not depend on $\mu$, $G^{\epsilon,1}(\mathbb{C}^-)$ is much smaller than, say, $G^{\epsilon,2}(\mathbb{C}^-)$ in a sense which we explain next.

Our proof of proposition 2.3 will establish a bijection between a basis for $G^{\epsilon,\mu}(\mathbb{C}^-)$ as a free abelian group and the isomorphism classes of self-dual simple finite-dimensional complex representations of a certain ring $P_\mu$. In other words, we associate to each[1] such isomorphism class $M$ one signature invariant

$$\sigma_{M,b} : C_{2q-1}(F_\mu) \to \mathbb{Z},$$

the composition of (13) with a projection onto a summand. The second subscript $b : M \to M^*$ is one of two possible forms and serves to specify a choice of sign for the signature. Note also that a simple representation $M$ must be 'algebraically integral' if the corresponding signature of an $F_\mu$-link is to be non-trivial.

The ring $P_\mu$, which is non-commutative when $\mu \geq 2$, was introduced by M.Farber [25]; we explain its role and outline the proof of proposition 2.3 in section 3 below.

We implicitly assume that all representations are finite-dimensional. In the case $\mu = 1$, $P_\mu$ is just the polynomial ring $\mathbb{Z}[s]$ with the involution $s \mapsto 1 - s$. By the fundamental theorem of algebra, the simple complex representations of $\mathbb{Z}[s]$ are precisely the one-dimensional representations where the action of $s$ is given by multiplication by a complex number $\nu$. The representation is self-dual if and only if $\bar{\nu} = 1 - \nu$, i.e. if and only if $\nu$ lies on the line[2] $\{\frac{1}{2} + bi \mid b \in \mathbb{R}\}$.

To obtain a corresponding geometric description when $\mu \geq 2$ we begin by observing that $P_\mu$ is the path ring of a certain quiver. A quiver, by definition, is a directed graph possibly with loops and multiple edges. The path ring of a quiver is free as a $\mathbb{Z}$-module with one basis element for each path in the quiver; the product of two paths is their concatenation if that

---

[1]Note that $\sigma_{M,b}$ is equal to the complex conjugate $\sigma_{\overline{M},b}$ so if $M \not\cong \overline{M}$ then we are really defining one signature to correspond to the pair $\{M, \overline{M}\}$.

[2]From the point of view of Blanchfield forms, one considers representations of $\mathbb{Z}[z, z^{-1}]$ in place of $\mathbb{Z}[s]$ where the involution on $\mathbb{Z}[z, z^{-1}]$ is $z \mapsto z^{-1}$. A simple complex representation is one-dimensional and the action of $z$ is multiplication by a complex number $\nu'$. The representation is self-dual if and only if $\bar{\nu'} = \nu'^{-1}$ so one signature is defined for each point on the unit circle. To relate signatures of Seifert and Blanchfield forms, the simple representation $s \mapsto \nu : \mathbb{C} \to \mathbb{C}$ gives rise to the presentation (6) with $\sigma = z\nu + (1-\nu)$. The action of $z$ on $\mathrm{Coker}(z\nu + (1-\nu))$ is multiplication by $\nu' = \nu^{-1}(\nu - 1) = 1 - \nu^{-1}$. The map $\nu \mapsto 1 - \nu^{-1}$ sends the line $\{\frac{1}{2} + ib \mid b \in \mathbb{R}\}$ to the unit circle. For further discussion of the relationship between Seifert forms and Blanchfield forms see for example Kearton [39], Levine [57] or Ranicki [84, Chapter 32].

makes sense or zero if it does not. See chapter 3, section 1.1 for further details.

A representation of a quiver is a collection of vector spaces, one for each vertex in the quiver, and a collection of linear maps, one for each arrow (i.e. directed edge). *The representation theory of a quiver is identical to the representation theory of its path ring.*

The particular quiver of which $P_\mu$ is the path ring is the 'complete quiver on $\mu$ vertices'. It contains $\mu^2$ arrows, one arrow for each ordered pair of vertices. For example, in the case $\mu = 2$ the quiver has the following appearance:

To describe geometrically the simple representations of $P_\mu$ one must first specify the dimensions of the vector spaces one wishes to associate to the vertices. These dimensions are usually written as a 'dimension vector' $\alpha = (\alpha_1, \cdots, \alpha_\mu)$. Using geometric invariant theory, the isomorphism classes of semisimple dimension vector $\alpha$ representations of $P_\mu$ correspond to the points on an (irreducible) affine algebraic variety $\mathcal{M}(P_\mu, \alpha)$. This variety is rarely smooth.

If there are any simple representations of dimension vector $\alpha$, then almost all are simple. More precisely the simple isomorphism classes are the points in a (Zariski) open smooth subvariety of $\mathcal{M}(P_\mu, \alpha)$, the top (Luna) stratum in a partition of $\mathcal{M}(P_\mu, \alpha)$ into 'representation types' (Le Bruyn and Procesi [**51**]). In the case $\mu = 1$, only the dimension vector $\alpha = (1)$ admits simple representations as we have discussed before. On the other hand, when $\mu \geq 2$, most dimension vectors admit simple representations (see lemma 6.3).

There is a restriction to impose, for we wish to isolate *self-dual* representations of $P_\mu$. The duality functor $M \mapsto M^*$ induces an involution on each variety $\mathcal{M}(P_\mu, \alpha)$ and the fixed point set $\overline{\mathcal{M}}(P_\mu, \alpha)$ of this involution turns out to be a real algebraic variety whose (real) dimension coincides with the (complex) dimension of $\mathcal{M}(P_\mu, \alpha)$. In summary,

PROPOSITION 2.4. *A basis of $G^{\epsilon,\mu}(\mathbb{C}^-)$ is in bijective correspondence with a Zariski open subset of an infinite disjoint union of (absolutely irreducible) real algebraic varieties $\bigsqcup_\alpha \overline{\mathcal{M}}(P_\mu, \alpha)$. The dimension of $\overline{\mathcal{M}}(P_\mu, \alpha)$ is $1 + \sum_{1 \leq i < j \leq \mu} 2\alpha_i \alpha_j$.*

In general an $F_\mu$-link may have non-zero signatures corresponding to several different simple, self-dual representations, and they need not lie on the same variety $\overline{\mathcal{M}}(P_\mu, \alpha)$. If $(L, \theta)$ is a split $F_\mu$-link then all its signatures lie on the one-dimensional varieties $\overline{\mathcal{M}}(P_\mu, \delta^i)$ where $\delta^i_j = \begin{cases} 1 & \text{if } i = j, \\ 0 & \text{otherwise.} \end{cases}$

Each of these varieties is a copy of the variety $\{\frac{1}{2} + bi \mid b \in \mathbb{R}\}$ of knot signatures.

**1.3. Character.** Since we are associating signatures to representations of $P_\mu$ we must learn to tell such representations apart. Moreover, if a signature $\sigma_{M,b} : G^{\epsilon,\mu}(\mathbb{Z}) \to \mathbb{Z}$ is to be non-zero then $M$ must be a summand of a representation which is induced up from an integral representation:

$$M \oplus M' \cong \mathbb{C} \otimes_{\mathbb{Z}} M_0.$$

We desire criteria to detect such 'algebraically integral' representations $M$.

The character of a representation

(14) $$\chi_M : P_\mu \to \mathbb{C}^-; \quad r \mapsto \mathrm{Trace}(\rho(r)),$$

which is an interpretation of M.Farber's 'trace invariant' (cf [25], [85]), is helpful in both respects. Firstly, a semisimple representation $M$ of any ring over any field of characteristic zero is determined by its character $\chi_M$ (see chapter 8). The representation is self-dual if and only if $\chi_M$ respects involutions.

Although $\chi$ may appear to be an infinite entity one can decide whether or not two representations of $P_\mu$ are isomorphic by a finite number of comparisons. Two $m$-dimensional complex representations $(M, \rho)$ and $(M', \rho')$ of $P_\mu$ are isomorphic if and only if $\chi_M(w) = \chi_{M'}(w)$ for all oriented cycles $w$ of length at most $m^2$ in the quiver (see Formanek [28]).

The character also detects algebraically integral representations; a complex representation (of any ring) is algebraically integral if and only if the character takes values in the integers of some algebraic number field.

## 2. Number Theory

Invariants which distinguish torsion elements of $C_{2q-1}(F_\mu)$ can be obtained by changing the ground ring not to $\mathbb{C}^-$ but to $\mathbb{Q}$.

KNOT THEORY EXAMPLE 2.5. The canonical map

$$G^{\epsilon,1}(\mathbb{Z}) \to G^{\epsilon,1}(\mathbb{Q})$$

is injective. Moreover, $G^{\epsilon,1}(\mathbb{Q})$ is isomorphic to an infinite direct sum of Witt groups of hermitian[3] forms over algebraic number fields (with non-trivial involution):

$$G^{\epsilon,1}(\mathbb{Q}) \cong \bigoplus_p W\left(\frac{\mathbb{Q}[s]}{p}\right)$$

There is one summand for each maximal ideal $(p) \triangleleft \mathbb{Q}[s]$ which is self-dual in the sense that $(p(s)) = (p(1-s)) \triangleleft \mathbb{Q}[s]$.

Hermitian forms over algebraic number fields were first classified by Landherr [49]. The Witt class of such a form is determined by the dimension

---

[3]To be precise, there is one summand $W^1\left(\frac{\mathbb{Q}[s]}{s-\frac{1}{2}}\right) = W^1(\mathbb{Q})$ which obviously has trivial involution. However the projection of $G^{\epsilon,1}(\mathbb{Z})$ onto this summand is zero because $\frac{1}{2}$ is not an integer - see chapter 11, section 3.

modulo 2, the signatures if any, and the discriminant. Up to sign, the discriminant of a form is just the determinant of any matrix which represents the form. Precise definitions can be found in section 2.1 of chapter 11.

PROPOSITION 2.6. *The canonical map*

$$G^{\epsilon,\mu}(\mathbb{Z}) \to G^{\epsilon,\mu}(\mathbb{Q})$$

*is injective. Moreover, $G^{\epsilon,\mu}(\mathbb{Q})$ is isomorphic to an infinite direct sum of Witt groups of finite-dimensional division $\mathbb{Q}$-algebras with involution:*

(15) $$G^{\epsilon,\mu}(\mathbb{Q}) \cong \bigoplus_M W^1(\text{End}(M)).$$

There is one summand in (15) for each isomorphism class $M$ of simple $\epsilon$-self-dual finite-dimensional rational representations of $P_\mu$ - see section 3 below. Note that the involution on the endomorphism algebra $\text{End}(M)$ and the isomorphism (15) both depend a choice of $\epsilon$-symmetric form $b: M \to M^*$.

Fortunately, complete invariants for the Witt groups of the division algebras $\text{End}(M)$ are available in the literature - for example [**1, 34, 47, 62, 64, 93**]. We summarize these invariants in theorem 11.5. For all classes of algebras but one, some combination of the following suffices to characterize the Witt class of a form: dimension modulo 2; signatures; discriminant; Hasse-Witt invariant. If $\text{End}(M)$ is a quaternion algebra with a 'non-standard' involution one requires a secondary invariant such as the Lewis $\theta$-invariant which is defined if the primary invariants vanish. We discuss all of these invariants in chapter 11 below. Using (15) they lead to an algorithm of to decide whether or not two $F_\mu$-links are cobordant when $q > 1$.

Proposition 2.6 also explains a qualitative difference between knot and $F_\mu$-link cobordism, the difference between Levine's theorem 2.1 and theorem A. Whereas knot theory leads us to study Witt groups of number fields $K_i$ which have non-trivial involution, it turns out that every finite-dimensional division algebra with involution appears as $\text{End}(M)$ in (15), with infinite multiplicity. In particular the Witt groups of all number fields with trivial involution are summands of $G^{\epsilon,\mu}(\mathbb{Q})$. One can find symmetric forms of order 8 in the Witt group of a number field if and only if $-1$ can be expressed as a sum of four squares, but not as a sum of fewer than four squares.

## 3. Defining Invariants

Let us outline the proofs of propositions 2.3 and 2.6. By studying the structure of $G^{\epsilon,\mu}(\mathbb{C}^-)$ we shall define signatures of $C_{2q-1}(F_\mu) \cong G^{\epsilon,\mu}(\mathbb{Z})$, one for each self-dual simple complex representation of $P_\mu$. These signatures will be defined in four steps. The first step is a 'change of variables'; the second 'devissage' step decomposes Seifert forms over $\mathbb{C}^-$ into simple constituents; the third step is a Morita equivalence which allows us to replace these simple constituents by their endomorphism rings; the fourth is Sylvester's theorem $W^1(\mathbb{C}^-) \cong \mathbb{Z}$.

Steps two and three are part of a general theory of hermitian forms formulated by Quebbemann, Scharlau and Schulte (see [80] and [93, Chapter 7]). We set out the details in chapters 4 and 5 below. The first three steps will be recycled, substituting $\mathbb{Q}$ for $\mathbb{C}^-$, when we define invariants to distinguish the torsion classes of $G^{\epsilon,\mu}(\mathbb{Z})$ and prove proposition 2.6.

**3.1. Sylvester's Theorem.** Since the notion of signature is so important in this work, we begin at the 'fourth step', the theory of hermitian forms over $\mathbb{C}^-$, however elementary the topic may be considered. Further details can be found in algebra textbooks such as Lang [50, p577].

Suppose $M$ is a finite-dimensional complex vector space and $\phi : M \to M^*$ is a non-singular hermitian form. With respect to any basis $x_1, \cdots x_m$, the form $\phi$ is represented by an invertible matrix $T_{ij} = \phi(x_i)(x_j)$ such that $\overline{T}^t = T$. An orthogonal basis can be chosen such that

$$\phi(x_i)(x_j) = \begin{cases} 1 \text{ or } -1 & \text{if } i = j \\ 0 & \text{if } i \neq j \end{cases}.$$

In other words, $T$ is congruent to a diagonal matrix with diagonal entries $\pm 1$. The number $m_+(\phi)$ of positive entries and the number $m_-(\phi)$ of negative entries do not depend on the choice of basis; they are well-defined invariants which determine $\phi$ uniquely up to isomorphism.

For the purposes of cobordism computations one aims to classify forms not up to isomorphism but up to Witt equivalence. The Witt group $W(\mathbb{C}^-)$ is by definition the semigroup of non-singular hermitian forms modulo the subsemigroup of metabolic forms. As we saw in section 4.2 of chapter 1 a form is metabolic if it is represented by a matrix $T = \begin{pmatrix} 0 & * \\ * & * \end{pmatrix}$ where each $*$ denotes some square matrix. In terms of invariants a non-singular hermitian form $\phi$ over $\mathbb{C}^-$ is metabolic if and only if $m_+(\phi) = m_-(\phi)$.

DEFINITION 2.7. The *signature* of a non-singular hermitian form is

$$\sigma(\phi) = m_+(\phi) - m_-(\phi).$$

THEOREM 2.8 (Sylvester). *The signature defines an isomorphism*

$$\sigma : W(\mathbb{C}^-) \xrightarrow{\cong} \mathbb{Z}.$$

**3.2. Change of Variables.** The first step in our definition of signature for a Seifert form is a 'symmetrization'. A more precise treatment than that in the present section can be found in chapter 3, section 4. Let us concentrate first on knot theory.

A Seifert form $\lambda : M \to M^*$, which is asymmetric, can be replaced by two entities: the $\epsilon$-symmetric form $\phi = \lambda + \epsilon\lambda^* : M \to M^*$ and an endomorphism $s = \phi^{-1}\lambda : M \to M$. The intersection form $\phi$ is intrinsic to the Seifert surface $V^{2q}$ while $s$ encodes homological information about the embedding of $V^{2q}$ in $S^{2q+1}$. The two are related by

(16) $$\phi(sx)(x') = \phi(x)((1-s)x') \text{ for all } x, x' \in M.$$

To rephrase things a little, a Seifert form $(M, \lambda)$ is replaced by an integral representation $\mathbb{Z}[s] \to \text{End}_{\mathbb{Z}} M$ of the polynomial ring $\mathbb{Z}[s]$ and an $\epsilon$-symmetric form $\phi : M \to M^*$ which respects the representation. The involution on $\mathbb{Z}[s]$ is given by $s \mapsto 1 - s$. This change of variables leads to an identity of Witt groups:

$$G^{\epsilon,1}(\mathbb{Z}) \cong W^{\epsilon}(\mathbb{Z}[s]\text{-}\mathbb{Z}) .$$

One can perform essentially the same trick for all $\mu \geq 1$. We start with slightly more data, a homomorphism $\lambda : M \to M^*$ and a system of $\mu$ projections $\pi_i : M \to M$. The intersection form $\phi = \lambda + \epsilon \lambda^*$ respects the projections, $\pi_i^* \phi = \phi \pi_i$, for there is no intersection between the homology classes of distinct components of a Seifert surface (the matrix $T$ represented $\phi$ in chapter 1, section 5.2). We define $s = \phi^{-1} \lambda : M \to M$ just as in knot theory, and observe that equation (16) still holds.

Let us again rephrase matters using a representation. One can adjoin to the polynomial ring $\mathbb{Z}[s]$ a system of orthogonal idempotents $\pi_1, \cdots, \pi_\mu$ obtaining a non-commutative ring

$$P_\mu \cong \mathbb{Z}\left\langle s, \pi_1, \cdots, \pi_\mu \;\middle|\; \pi_i^2 = \pi_i; \pi_i \pi_j = 0 \text{ for } i \neq j; \sum_{i=1}^{\mu} \pi_i = 1 \right\rangle.$$

$$\cong \mathbb{Z}[s] *_{\mathbb{Z}} \left(\prod_\mu \mathbb{Z}\right)$$

With the appropriate involution on $P_\mu$,

$$s \mapsto 1 - s; \quad \pi_i \mapsto \pi_i \text{ for } 1 \leq i \leq \mu,$$

a $\mu$-component Seifert form $(M, \lambda, \{\pi_i\})$ becomes an integral representation $P_\mu \to \text{End}_{\mathbb{Z}} M$ together with a form $\phi : M \to M^*$ which respects the representation (cf Farber [**25**]). The corresponding isomorphism of Witt groups is the following:

$$\kappa : G^{\epsilon,\mu}(\mathbb{Z}) \xrightarrow{\cong} W^{\epsilon}(P_\mu\text{-}\mathbb{Z})$$

where $W^{\epsilon}(P_\mu\text{-}\mathbb{Z})$ is the Witt group of triples

$$(M, \rho : P_\mu \to \text{End}_{\mathbb{Z}} M, \phi : M \to M^*).$$

**3.3. Devissage.** The second step in the definition of signature is a hermitian version of the Jordan-Hölder theorem. Further details can be found in chapter 5. The Jordan-Hölder theorem says that in a context where representations admit finite composition series

$$0 = N_0 \subset N_1 \subset \cdots \subset N_l = M$$

such that each subfactor $N_i/N_{i-1}$ is simple, the subfactors of any two composition series for the same representation are isomorphic, possibly after reordering.

The hermitian version we need identifies an element of the Witt group $W^\epsilon(P_\mu\text{-}\mathbb{C}^-)$ with a direct sum of Witt classes of forms defined over simple representations, the subfactors. Of course, a simple representation $M$ must be self-dual if there is to be a non-singular form $\phi : M \to M^*$. Indeed, it turns out that any non-self-dual subfactors 'cancel out' in pairs. The appropriate uniqueness theorem is expressed by the canonical isomorphism

$$(17) \qquad W^\epsilon(P_\mu\text{-}\mathbb{C}^-) \cong \bigoplus_M W^\epsilon_M(P_\mu\text{-}\mathbb{C}^-)$$

with one summand for each isomorphism class of simple self-dual complex representations $M$ of $P_\mu$. The summand $W^\epsilon_M(P_\mu\text{-}\mathbb{C}^-)$ is by definition the Witt group of $\epsilon$-hermitian forms

$$\phi : M \oplus \cdots \oplus M \to (M \oplus \cdots \oplus M)^* .$$

over $M$-isotypic representations $M \oplus \cdots \oplus M$.

NOTATION 2.9. Let us denote by $p_M$ the canonical projection

$$W^\epsilon(P_\mu\text{-}\mathbb{C}^-) \twoheadrightarrow W^\epsilon_M(P_\mu\text{-}\mathbb{C}^-).$$

A complex representation of $P_\mu$ can of course be expressed as a module over $\mathbb{C}^- \otimes_\mathbb{Z} P_\mu$. In knot theory, $P_1 = \mathbb{Z}[s]$ and $\mathbb{C} \otimes_\mathbb{Z} \mathbb{Z}[s] \cong \mathbb{C}[s]$ is a principal ideal domain. J. Milnor [69] used the structure theorem for modules over principal ideal domains to prove a version of (17) when $\mu = 1$.

**3.4. Hermitian Morita Equivalence.** The third step, which we treat more carefully in chapter 4, is the following isomorphism

$$W^\epsilon_M(P_\mu\text{-}\mathbb{C}^-) \cong W^1(\text{End}(M))$$

which replaces an isotypic representation $M \oplus \cdots \oplus M$ on the left hand side by

$$\text{Hom}(M, M \oplus \cdots \oplus M)$$

on the right hand side. The latter admits a natural 'composition' action by the endomorphism ring $\text{End}(M)$.

Since $M$ is a simple representation, the endomorphism ring $\text{End}(M)$ is a division ring (Schur's lemma). All the representations with which we are concerned are finite-dimensional so $\text{End}(M)$ is also finite-dimensional and is therefore isomorphic to $\mathbb{C}^-$.

NOTATION 2.10. Let us denote the Morita equivalence by

$$(18) \qquad \Theta_{M,b} : W^\epsilon_M(P_\mu\text{-}\mathbb{C}^-) \xrightarrow{\cong} W^1(\mathbb{C}^-).$$

Unlike (17) above, the isomorphism $\Theta_{M,b}$ depends on a choice of $\epsilon$-hermitian form $b : M \to M^*$. There are precisely two 1-dimensional forms over $\mathbb{C}^-$, namely $\langle 1 \rangle$ and $\langle -1 \rangle$. It follows that there are precisely two choices for $b$; in other words one must make a choice of sign in the definition of each signature.

To summarize all four steps we have the following definition:

DEFINITION 2.11. If $M$ is a finite-dimensional simple complex representation of $P_\mu$ and $\epsilon b^* = b : M \to M^*$ is a non-singular $\epsilon$-hermitian form then the signature $\sigma_{M,b}$ is the composite

$$G^{\epsilon,\mu}(\mathbb{Z}) \to G^{\epsilon,\mu}(\mathbb{C}^-) \xrightarrow{\kappa} W^\epsilon(P_\mu\text{-}\mathbb{C}^-) \xrightarrow{p_M} W^\epsilon_M(P_\mu\text{-}\mathbb{C}^-) \xrightarrow{\Theta_{M,b}} W^\epsilon(\mathbb{C}^-) \xrightarrow{\sigma} \mathbb{Z}$$

We define the signature $\sigma_{M,b}(L,\theta)$ of an $F_\mu$-link to be the corresponding signature of any Seifert form for $(L,\theta)$.

### 3.5. Number Theory Invariants.

To distinguish torsion classes in $G^{\epsilon,\mu}(\mathbb{Z})$, we apply the first step, the 'change of variables', exactly as before. We replace $\mathbb{C}^-$ by $\mathbb{Q}$ and perform devissage and hermitian Morita, obtaining

$$G^{\epsilon,\mu}(\mathbb{Z}) \to G^{\epsilon,\mu}(\mathbb{Q}) \xrightarrow{\kappa} W^\epsilon(P_\mu\text{-}\mathbb{Q}) \cong \bigoplus_M W^\epsilon_M(P_\mu\text{-}\mathbb{Q}) \cong \bigoplus_M W^1(\text{End}(M)).$$

The map $\kappa$ is an isomorphism (lemma 3.31) so $G^{\epsilon,\mu}(\mathbb{Q}) \cong \bigoplus_M W^1(\text{End}(M))$.

We denote the projection onto the $M$th component

$$p_M : W^\epsilon(P_\mu\text{-}\mathbb{Q}) \twoheadrightarrow W^\epsilon_M(P_\mu\text{-}\mathbb{Q})$$

and the Morita equivalence

$$\Theta_{M,b} : W^\epsilon_M(P_\mu\text{-}\mathbb{Q}) \xrightarrow{\cong} W^1(\text{End}(M)).$$

A straightforward lemma - lemma 11.1 - shows that the canonical map $G^{\epsilon,\mu}(\mathbb{Z}) \to G^{\epsilon,\mu}(\mathbb{Q})$ is injective, completing the proof of proposition 2.6. We can finally state our main result:

THEOREM B. *Suppose $(L^0,\theta^0)$ and $(L^1,\theta^1)$ are $(2q-1)$-dimensional $F_\mu$-links, where $q > 1$. For $i = 0,1$ let $V^i$ denote a $\mu$-component Seifert surface for $(L^i,\theta^i)$ and let $N^i = H_q(V^i)/\text{torsion}$. Let $\lambda^i : N^i \to (N^i)^*$ be the Seifert form corresponding to $V^i$.*

*The $F_\mu$-links $(L^0,\theta^0)$ and $(L^1,\theta^1)$ are cobordant if and only if for each finite-dimensional $\epsilon$-self-dual simple rational representation $M$ of $P_\mu$, the dimension modulo 2, the signatures, the discriminant, the Hasse-Witt invariant and the Lewis $\theta$-invariant of*

$$\Theta_{M,b}\, p_M\, \kappa\, [\mathbb{Q} \otimes_\mathbb{Z} (N^0 \oplus N^1, \lambda^0 \oplus -\lambda^1)] \in W^1(\text{End}(M))$$

*are trivial (if defined).*

Note that one need compute only a finite number of invariants to establish that two given $F_\mu$-links are cobordant. Note also that if all the invariants vanish for some choice of $b : M \to M^*$ then they vanish for all possible choices.

## 4. Chapter Summaries

In chapter 3 we collect some definitions and basic properties of quivers, Grothendieck groups, Witt groups and Seifert forms. We treat the 'change of variables' outlined in section 3.2.

Chapters 4 and 5 develop the machinery required to decompose $G^{\epsilon,\mu}(\mathbb{C}^-)$ and $G^{\epsilon,\mu}(\mathbb{Q})$ as direct sums of Witt groups of division rings. Chapter 4 concerns Morita equivalence, the 'third step' described in section 3.4 above. We work in an arbitrary additive category with hermitian structure and split idempotents. Chapter 5 treats the 'second step', the devissage technique of section 3.3 above. Here, we work in any abelian category which has a hermitian structure and in which objects have finite composition series.

Chapter 6 describes the variety structures for the signatures, proving proposition 2.4.

Chapter 7 gives an algebraic proof, adapted from work of Scharlau, that the order of every element in the kernel of the canonical map $G^{\epsilon,\mu}(\mathbb{Q}) \to G^{\epsilon,\mu}(\mathbb{C}^-)$ is a power of 2. This implies that the signatures in definition 2.11 are a complete set of torsion-free invariants. In fact we prove a generalization which applies to arbitrary fields.

Chapter 8 contains a proof that a semisimple representation is determined up to isomorphism by its character. Chapter 9 uses the character to give criteria for a semisimple representation to be a summand of a rational or integral representation.

In chapter 10 we compute explicitly two examples of the real varieties $\overline{\mathcal{M}}(P_2, \alpha)$ and the Zariski open subsets of simple self-dual representations. One signature of $C_{2q-1}(F_2)$ has been defined for each algebraic integer, or complex conjugate pair of algebraic integers, in these open subsets.

Chapter 11 describes invariants of Witt groups of finite-dimensional division $\mathbb{Q}$-algebras, completing the proofs of propositions 2.3 and 2.6 and of theorem B. We go on to discuss the localization exact sequence

$$0 \to W^\epsilon(P_\mu\text{-}\mathbb{Z}) \to W^\epsilon_\mathbb{Z}(P_\mu\text{-}\mathbb{Q}) \to W^\epsilon(P_\mu\text{-}\mathbb{Q}/\mathbb{Z}) \to \cdots$$

Following the work of Stolzfus [97] on knot cobordism, one could use this exact sequence in an attempt to describe $C_{2q-1}(F_\mu)$ with greater precision, to characterize the subgroup $G^{\epsilon,\mu}(\mathbb{Z}) \subset G^{\epsilon,\mu}(\mathbb{Q})$ in terms of invariants.

Finally, in chapter 12 we construct examples of integral representations $M$ of $P_\mu$ such that the endomorphism ring of $\mathbb{Q} \otimes_\mathbb{Z} M$ is *any* prescribed finite-dimensional division algebra with involution. Since every class of division algebras arises, all the invariants in the previous chapter are warranted. Chapter 12 concludes with a proof of theorem A.

CHAPTER 3

# Preliminaries

In this chapter we recall the definitions of the Grothendieck group and Witt group of an associative ring and, more generally, of an abstract category with sufficient structure. Our main example is the category of finitely generated projective representations of an associative ring $R$ with involution over another associative ring $A$. In particular, we are concerned with representations of the particular ring $P_\mu$ over $\mathbb{Z}$ or over a field.

In section 4 we define $\mu$-component Seifert forms and prove that the Witt group of such forms is isomorphic to the Witt group of representations of $P_\mu$.

## 1. Representations

Let $A$ and $R$ denote associative rings with identity. A *representation* $(M, \rho)$ of $R$ over $A$ will be a ring homomorphism $\rho : R \to \mathrm{End}_A(M)$ where $M$ is a finitely generated projective $A$-module. We often write $M$ in place of $(M, \rho)$.

Let $(R\text{–}A)$-Proj denote the category of representations of $R$ over $A$. By definition a morphism $\theta : (M, \rho) \to (M', \rho')$ in $(R\text{–}A)$-Proj is an $A$-module map $\theta : M \to M'$ such that $\rho'(r)\theta = \theta\rho(r)$ for all $r \in R$.

A ring homomorphism $A \to B$ induces a functor

$$(R\text{–}A)\text{-Proj} \to (R\text{–}B)\text{-Proj}$$

as follows: If $(M, \rho)$ is a representation of $R$ over $A$ then $(B \otimes_A M, \rho_B)$ is a representation over $B$ where

$$\rho_B : R \to \mathrm{End}_B(B \otimes_A M)$$
$$r \mapsto 1 \otimes \rho(r).$$

In our application to boundary links, $A$ will be a field $A = k$ or the integers $A = \mathbb{Z}$ and $R$ will be a path ring for a quiver:

**1.1. Representations of Quivers.** For the convenience of the reader we summarize the basic definitions and properties of quivers, paraphrasing Benson [6, p99]. Further background can be found in the book by Auslander, Reiten and Smalø [2, §3.1].

DEFINITION 3.1. A quiver $Q$ is a directed graph, possibly with loops and multiple directed edges (arrows). It has a finite set $Q_0$ of vertices and a set $Q_1$ of arrows. Each arrow $e \in Q_1$ has a head $h(e) \in Q_0$ and a tail $t(e) \in Q_0$.

A representation of a quiver will be a system $\{M_x\}_{x\in Q_0}$ of projective $A$-modules together with a system of homomorphisms
$$\{f_e : M_{t(e)} \to M_{h(e)}\}_{e\in Q_1}.$$
If $A$ is a field, say $A = k$, then each representation $(M, \rho)$ has a *dimension vector*
$$\alpha = (\dim_k M_x)_{x\in Q_0} \in \mathbb{Z}_{\geq 0}^{Q_0}.$$
Over any $A$, a representation of a quiver can be interpreted as a representation $\rho : \mathbb{Z}Q \to \text{End}_A(M)$ of the path ring $\mathbb{Z}Q$ which we define next.

**1.2. The Path Ring of a Quiver.** The path ring $\mathbb{Z}Q$ is a free $\mathbb{Z}$-module with one basis element for each non-trivial path
$$p = \left(\bullet \xrightarrow{e_1} \bullet \xrightarrow{e_2} \cdots \xrightarrow{e_n} \bullet\right)$$
in $Q$ and one basis element $\pi_x$ corresponding to the trivial path at each vertex $x \in Q_0$. If $p$ is non-trivial we write $h(p) = h(e_n)$ and $t(p) = t(e_1)$ whereas for trivial paths we set $h(\pi_x) = t(\pi_x) = x$.

If $h(q) = t(p)$ then the product $p.q$ is by definition 'path $q$ followed by path $p$', the composite of the paths in reverse order. If $h(q) \neq t(p)$ then one defines $p.q = 0$. Note that the trivial paths $\{\pi_x\}_{x\in Q_0}$ form a system of orthogonal idempotents in $\mathbb{Z}Q$
$$\sum_{x\in Q_0} \pi_x = 1; \quad \pi_x \pi_y = \begin{cases} \pi_x & \text{if } x = y \\ 0 & \text{if } x \neq y. \end{cases}$$

KNOT THEORY EXAMPLE 3.2. Let $Q$ be the quiver with one vertex and one arrow. The path ring $\mathbb{Z}Q$ is (isomorphic to) the ring of polynomials $\mathbb{Z}[s]$ in an indeterminate $s$.

A representation $(\{M_x\}_{x\in Q_0}, \{f_e\}_{e\in Q_1})$ of any quiver $Q$ can be rewritten
$$\left(\bigoplus_{x\in Q_0} M_x, \rho\right)$$
where $\rho : \mathbb{Z}Q \to \text{End}(\bigoplus_{x\in Q_0} M_x)$ and
$$\rho\left(\bullet \xrightarrow{e_1} \bullet \xrightarrow{e_2} \cdots \xrightarrow{e_n} \bullet\right)$$
is the composite of the corresponding maps
$$\bigoplus M_x \twoheadrightarrow M_{t(e_1)} \xrightarrow{f_{e_1}} \cdots \xrightarrow{f_{e_n}} M_{h(e_n)} \hookrightarrow \bigoplus M_x.$$
Inversely, given a representation $(M, \rho)$ of $\mathbb{Z}Q$ one may define $M_x = \rho(\pi_x)M$ and let $f_e : M_{t(e)} \to M_{h(e)}$ be the restriction of $\rho(e)$. Note that $\rho(e)$ maps $M_{t(e)}$ to $M_{h(e)}$ since $\rho(e) = \rho(e_{h(e)}e) = \rho(ee_{t(e)})$. Henceforth we shall freely interchange the symbols $Q$ and $\mathbb{Z}Q$, writing for example $(Q\text{–}A)\text{-Proj} = (\mathbb{Z}Q\text{–}A)\text{-Proj}$.

**1.3. Boundary Links.** We require the following quivers in the study of boundary links (compare Farber [**25**], [**24**]) :

DEFINITION 3.3. Let $P_\mu$ denote the quiver with $\mu$ vertices $x_1, \cdots, x_\mu$ and precisely one edge joining vertex $x_i$ to vertex $x_j$ for each ordered pair $(i, j)$ with $1 \leq i, j \leq \mu$.

For example, in the case $\mu = 2$ the quiver $P_2$ appears as follows:

$$\circlearrowleft \bullet \rightleftarrows \bullet \circlearrowright$$

As usual, $P_\mu$ will denote both the quiver and its path ring.

LEMMA 3.4. *The path ring $P_\mu$ is isomorphic to the free product of a polynomial ring $\mathbb{Z}[s]$ by a product of $\mu$ copies of $\mathbb{Z}$. In symbols*

$$P_\mu \cong \mathbb{Z}[s] *_\mathbb{Z} \left( \prod_\mu \mathbb{Z} \right)$$

$$\cong \mathbb{Z} \left\langle s, \pi_1, \cdots, \pi_\mu \,\middle|\, \sum_{i=1}^\mu \pi_i = 1, \pi_i^2 = \pi_i, \pi_i \pi_j = 0 \text{ for } 1 \leq i, j \leq \mu \right\rangle.$$

PROOF. The indeterminate $s$ corresponds to the sum of all the paths of length one. The idempotents $\pi_i$ correspond to the trivial paths $\pi_{x_i}$. □

A representation of $P_\mu$ is to be thought of as the middle homology $H_q(V; A)$ of a $\mu$-component embedded Seifert surface $V^{2q}$ - see chapter 1, section 5.2 above and lemma 3.31 below.

## 2. Grothendieck Groups

**2.1. Rings.** Let $A$ be a ring, assumed to be associative and to contain a 1. Let $A$-Proj denote the category of projective left $A$-modules.

DEFINITION 3.5. $K_0(A)$ is the abelian group with one generator $[M]$ for each isomorphism class of finitely generated projective $A$-modules and one relation $[M'] = [M] + [M'']$ for each identity $M' \cong M \oplus M''$.

A ring homomorphism $A \to B$ induces an additive functor

$$A\text{-Proj} \to B\text{-Proj}; \quad M \mapsto B \otimes_A M$$

and therefore a group homomorphism $K_0(A) \to K_0(B)$.

**2.2. Categories.** More generally, one can define the Grothendieck group $K_0(\mathcal{C})$ of any category $\mathcal{C}$ which has a small skeleton and in which exact sequences are defined (see Rosenberg [**88**, ch3]). However, the following is general enough for our purposes:

DEFINITION 3.6. Suppose $\mathcal{C}$ is a full subcategory of an abelian category $\mathcal{A}$. The Grothendieck group $K_0(\mathcal{C})$ is the abelian group with one generator $[M]$ for each isomorphism class of objects in $\mathcal{C}$ and one relation $[M'] = [M] + [M'']$ for each exact sequence $0 \to M \to M' \to M'' \to 0$ in $\mathcal{C}$.

$K_0$ is functorial; if $F : \mathcal{C} \to \mathcal{D}$ is an exact functor, i.e. $F$ preserves exact sequences, then there is an induced homomorphism $F : K_0(\mathcal{C}) \to K_0(\mathcal{D})$ of abelian groups.

The most important example here is the category $\mathcal{C} = (R\text{–}A)$-Proj of representations defined in section 1: A sequence of objects and morphisms

(19) $$0 \to (M, \rho) \xrightarrow{\theta} (M', \rho') \xrightarrow{\theta'} (M'', \rho'') \to 0$$

is said to be exact if the underlying sequence $0 \to M \xrightarrow{\theta} M' \xrightarrow{\theta'} M'' \to 0$ is exact. $(R\text{–}A)$-Proj is a full subcategory of the abelian category $(R\text{–}A)$-Mod of representations of $R$ by arbitrary $A$-modules. We write

$$K_0(R\text{–}A) = K_0((R\text{–}A)\text{-Proj}).$$

For example, since an exact sequence of projective modules splits, we have

$$K_0(\mathbb{Z}\text{–}A) = K_0(A\text{-Proj}) = K_0(A).$$

On the other hand, if $R$ is any ring and $A = k$ is a field then $(R\text{–}k)$-Proj is an abelian category and the Jordan-Hölder theorem implies that $K_0(R\text{–}k)$ is a free abelian group with one generator for each isomorphism class of simple objects.

If $j : A \to B$ is a ring homomorphism then there is induced an exact functor $(R\text{–}A)$-Proj $\to (R\text{–}B)$-Proj (see section 1); for in any exact sequence (19) the underlying exact sequence of projective modules is split. Thus $j$ induces a group homomorphism $K_0(R\text{–}A) \to K_0(R\text{–}B)$.

## 3. Witt Groups

### 3.1. Hermitian Forms over a Ring.
Let $A$ be an associative ring.

DEFINITION 3.7. The opposite ring $A^o$ is identical to $A$ as an additive group but multiplication in $A^o$ is reversed: $a \circ a' = a'a$.

DEFINITION 3.8. An involution on $A$ is an isomorphism $A \to A^o$, usually denoted $a \mapsto \bar{a}$, such that $a = \bar{\bar{a}}$ for all $a \in A$.

An involution on $A$ induces an equivalence of categories Proj-$A \to$ $A$-Proj; a right $A$-module $M$ becomes a left $A$-module with $a.x = x\bar{a}$ for all $a \in A$ and $x \in M$. In particular, if $M \in A$-Proj then the dual module $M^* = \text{Hom}_A(M, A)$ is again a finitely generated projective left $A$-module with $(a.\xi)(x) = \xi(x)\bar{a}$ for $a \in A$, $\xi \in M^*$ and $x \in M$. We identify $M$ with $M^{**}$ via the natural isomorphism $M \to M^{**}; x \mapsto (\xi \mapsto \xi(x))$.

Let $\epsilon = 1$ or $-1$. An $\epsilon$-hermitian form is a pair $(M, \phi)$ where $M \in A$-Proj and $\phi : M \to M^*$ satisfies $\phi^* = \epsilon\phi$. If $\phi$ is an isomorphism then $(M, \phi)$ is said to be *non-singular*. The category of non-singular $\epsilon$-hermitian forms is denoted $H^\epsilon(A)$; a morphism $f : (M, \phi) \to (M', \phi')$ in $H^\epsilon(A)$ is a map $f : M \to M'$ such that $\phi = f^*\phi' f$.

If $(M, \phi)$ is an $\epsilon$-hermitian form and $j : L \hookrightarrow M$ is the inclusion of a summand, one defines

$$L^\perp := \text{Ker}\,(j^*\phi : M \to L^*).$$

If $L = L^\perp$ then $L$ is called a *lagrangian* or *metabolizer* and $(M,\phi)$ is called *metabolic*. Equivalently, $(M,\phi)$ is metabolic if and only if

$$(M,\phi) \cong \left( L \oplus L^*, \begin{pmatrix} 0 & 1 \\ \epsilon & b \end{pmatrix} : L \oplus L^* \to L^* \oplus L \right)$$

for some $L \in A$-Proj and some $\epsilon$-hermitian $b$.

DEFINITION 3.9. The Witt group $W^\epsilon(A)$ is the abelian group with one generator $[M,\phi]$ for each isomorphism class of non-singular $\epsilon$-hermitian forms in $H^\epsilon(A)$ subject to relations

$$\begin{cases} [M',\phi'] = [M,\phi] + [M'',\phi''], & \text{if } (M',\phi') \cong (M,\phi) \oplus (M'',\phi'') \\ [M,\phi] = 0, & \text{if } (M,\phi) \text{ is metabolic.} \end{cases}$$

Thus two forms represent the same Witt class $[M,\phi] = [M',\phi']$ if and only if there exist metabolic forms $(H,\eta)$ and $(H',\eta')$ such that

$$(M \oplus H, \phi \oplus \eta) \cong (M' \oplus H', \phi' \oplus \eta').$$

REMARK 3.10. An $\epsilon$-hermitian form $(M,\phi)$ is *hyperbolic* if there exists $L \in A$-Proj such that

$$(M,\phi) \cong \left( L \oplus L^*, \begin{pmatrix} 0 & 1 \\ \epsilon & 0 \end{pmatrix} : L \oplus L^* \to (L \oplus L^*)^* \right).$$

Although metabolic forms are not in general hyperbolic, the same group $W^\epsilon(A)$ is obtained if one substitutes the word 'hyperbolic' for 'metabolic' in definition 3.9 [**93**, p23].

REMARK 3.11. The Witt group $W^\epsilon(A)$ is isomorphic to a symmetric $L$-group $L^0_p(A,\epsilon)$ (Ranicki [**81**]). If there exists a central element $a \in A$ such that $a + \bar{a} = 1$ then $W^\epsilon(A)$ is also isomorphic to the quadratic $L$-group $L^p_{2q}(A, \epsilon(-1)^q)$ for all $q$.

EXAMPLE 3.12. For any ring $A$ the product $A \times A^o$ admits a transposition involution $\overline{(a,a')} = (a',a)$, for which $W^\epsilon(A \times A^o) = 0$.

PROOF. Any $\epsilon$-hermitian form $(M,\phi)$ over $A \times A^o$ has metabolizer $(1,0)M$. □

EXAMPLE 3.13. Let $\mathbb{C}^-$ denote the field of complex numbers with involution given by complex conjugation. Let $\mathbb{C}^+$ denote the same field with trivial involution. Then $W^\epsilon(\mathbb{C}^-) \cong \mathbb{Z}$ and $W^\epsilon(\mathbb{C}^+) \cong \mathbb{Z}/2\mathbb{Z}$.

REMARK 3.14. If $A$ is a commutative ring then the group $W^1(A)$ is also a commutative ring with multiplication given by the tensor product

$$(M,\phi) \otimes (M',\phi') = (M \otimes_A M', \phi \otimes \phi')$$

where $(\phi \otimes \phi')(x_1 \otimes x'_1)(x_2 \otimes x'_2) = \phi(x_1)(x_2)\phi'(x'_1)(x'_2)$ for $x_1, x_2 \in M$ and $x'_1, x'_2 \in M'$.

## 3.2. Change of Rings.
A homomorphism $j : A \to B$ of rings with involution induces a functor

$$H^\epsilon(A) \to H^\epsilon(B)$$
$$(M, \phi) \mapsto (B \otimes_A M, \phi_B)$$

where

$$\phi_B : B \otimes_A M \to \operatorname{Hom}_B(B \otimes_A M, B);$$
$$b \otimes x \mapsto (b' \otimes x' \mapsto b' j(\phi(x)(x')) \bar{b}).$$

for all $b, b' \in B$ and all $x, x' \in M$.

If $(M, \phi)$ is metabolic then $(B \otimes_A M, \phi_B)$ is again metabolic so $j$ induces a group homomorphism $W(A) \to W(B)$. If $A$ and $B$ are commutative then $j : W(A) \to W(B)$ is a ring homomorphism.

## 3.3. Hermitian Categories.
Quebbemann, Scharlau and Schulte formulated a more general theory of quadratic and hermitian forms over a wider class of categories [80]. We summarize next the definitions we require.

Suppose $\mathcal{C}$ is an additive category (see Bass [5, pp12-20]).

DEFINITION 3.15. A *duality functor* $* : \mathcal{C} \to \mathcal{C}$ is an additive contravariant functor together with a natural isomorphism $(i_M)_{M \in \mathcal{C}} : \text{id} \to **$ such that $i_M^* i_{M^*} = id_{M^*}$ for all $M \in \mathcal{C}$. We often omit $i$ identifying $M$ with $M^{**}$. A triple $(\mathcal{C}, *, i)$ is called a *hermitian category*, and is usually written $\mathcal{C}$ for brevity.

Our main examples of hermitian categories are $A$-Proj and $(R\text{–}A)$-Proj:

EXAMPLE 3.16. The category $A$-Proj admits the duality functor $M \mapsto M^* = \operatorname{Hom}_A(M, A)$. The category $(R\text{–}A)$-Proj of representations of a ring $R$ by f.g. projective $A$-modules admits the duality functor $(M, \rho) \mapsto (M, \rho)^* = (M^*, \rho^*)$ where

$$\rho^*(r)(\xi) = (x \mapsto \xi(\bar{r}.x))$$

for $r \in R$, $\xi \in M^*$ and $x \in M$.

DEFINITION 3.17. Let $\epsilon = +1$ or $-1$. An $\epsilon$-hermitian form over $\mathcal{C}$ is by definition a pair $(M, \phi)$ where $M$ is an object of $\mathcal{C}$ and $\phi : M \to M^*$ satisfies $\phi^* i_M = \epsilon \phi$. $(M, \phi)$ is *non-singular* if $\phi$ is an isomorphism.

DEFINITION 3.18. An object $M$ is called *self-dual* if $M \cong M^*$ and is called $\epsilon$-self-dual if there exists a non-singular $\epsilon$-hermitian form $(M, \phi)$.

In many cases of interest, self-dual objects are both 1-self-dual and $(-1)$-self-dual - see lemmas 5.5 and 5.6 below.

The category of non-singular $\epsilon$-hermitian forms is denoted $H^\epsilon(\mathcal{C})$.

## 3.4. Change of Hermitian Category.
Hermitian categories are themselves the objects of a category. We define next the morphisms between hermitian categories $\mathcal{C}$ and $\mathcal{D}$.

## 3. WITT GROUPS

DEFINITION 3.19. A *duality preserving functor* $\mathcal{C} \to \mathcal{D}$ is a triple $(F, \Phi, \eta)$ where $F : \mathcal{C} \to \mathcal{D}$ is an additive functor, $\{\Phi_M\}_{M \in \mathcal{C}} : F(\_^*) \to F(\_)^*$ is a natural isomorphism, $\eta = 1$ or $-1$ and

$$\Phi_M^* i_{F(M)} = \eta \Phi_{M^*} F(i_M) : F(M) \to F(M^*)^* \tag{20}$$

for all $M \in \mathcal{C}$.

A duality preserving functor $(F, \Phi, \eta)$ is an equivalence of hermitian categories if and only if $F$ is an equivalence of categories - see proposition II.7 of appendix II. Precise definitions of composition and equivalence of duality preserving functors are also given in appendix II.

EXAMPLE 3.20. Suppose as in section 3.2 above that $j : A \to B$ is a ring homomorphism and $B \otimes_A \_ : A\text{-Proj} \to B\text{-Proj}$ is the induced functor. There is a duality preserving functor $(B \otimes_A \_, \Phi, 1) : A\text{-Proj} \to B\text{-Proj}$, where $\Phi$ is the natural isomorphism

$$\Phi_M : B \otimes_A M^* = B \otimes_A \operatorname{Hom}_A(M, A) \to \operatorname{Hom}_B(B \otimes_A M, B) = (B \otimes_A M)^*$$

$$b \otimes \xi \mapsto (b' \otimes x \mapsto b'\xi(x)\bar{b}).$$

In the slightly more general setting of representation categories, essentially the same triple $(B \otimes_A \_, \Phi, 1) : (R\text{–}A)\text{-Proj} \to (R\text{–}B)\text{-Proj}$ is a duality preserving functor.

An example in which $\eta$ can be $-1$ will appear in the next chapter which concerns hermitian Morita equivalence.

LEMMA 3.21. *A duality preserving functor* $(F, \Phi, \eta) : \mathcal{C} \to \mathcal{D}$ *induces an additive functor between categories of non-singular hermitian forms:*

$$H^\epsilon(\mathcal{C}) \to H^{\epsilon\eta}(\mathcal{D})$$
$$(M, \phi) \mapsto (F(M), \Phi_M F(\phi)).$$

PROOF. Suppose $(M, \phi) \in H^\epsilon(\mathcal{C})$ is an $\epsilon$-hermitian form, so that $\phi^* i_M = \epsilon \phi$. To show $(F(M), \Phi_M F(\phi)) \in H^{\epsilon\eta}(\mathcal{D})$ we need $(\Phi_M F(\phi))^* i_{F(M)} = \epsilon\eta \Phi_M F(\phi)$. Indeed,

$$\begin{aligned}(\Phi_M F(\phi))^* i_{F(M)} &= F(\phi)^* \Phi_M^* i_{F(M)} \\ &= \eta F(\phi)^* \Phi_{M^*} F(i_M) \quad \text{by equation (20)} \\ &= \eta \Phi_M F(\phi^*) F(i_M) \quad \text{by naturality of } \Phi \\ &= \eta \Phi_M F(\phi^* i_M) \\ &= \epsilon\eta \Phi_M F(\phi).\end{aligned}$$

$\square$

**3.5. The Witt Group of a Category.** Suppose now that $\mathcal{C}$ is a hermitian category which is a full subcategory of an abelian category $\mathcal{A}$. Suppose further that $\mathcal{C}$ is admissible in $\mathcal{A}$, i.e. that if $0 \to M \to M' \to M'' \to 0$ is an exact sequence in $\mathcal{A}$ and $M'$ and $M''$ are in $\mathcal{C}$ then $M \in \mathcal{C}$.

A subobject $j : L \hookrightarrow M$ is called admissible in $\mathcal{C}$, if $\operatorname{Coker}(j : L \hookrightarrow M) \in \mathcal{C}$. In the case $\mathcal{C} = A\text{-Proj}$ of section 3.9, admissible subobjects are precisely direct summands.

If $L$ is admissible one defines
$$L^\perp = \text{Ker}\,(j^*\phi : M \to L^*).$$
which is an object of $\mathcal{C}$. We denote by $j^\perp$ the inclusion $L^\perp \hookrightarrow M$. If $L \subset L^\perp$ then $L$ is called a *sublagrangian* of $(M, \phi)$ while if $L = L^\perp$ then one says $L$ is a *lagrangian* or *metabolizer* for $(M, \phi)$ and $(M, \phi)$ is *metabolic*.

The definition of the Witt group of $\mathcal{C}$ is very similar to the Witt group of a ring (see definition 3.9 above):

DEFINITION 3.22. The Witt group $W^\epsilon(\mathcal{C})$ is the abelian group with one generator $[M, \phi]$ for each isomorphism class of non-singular $\epsilon$-hermitian forms $(M, \phi) \in H^\epsilon(\mathcal{C})$ subject to relations
$$\begin{cases} [M', \phi'] = [M, \phi] + [M'', \phi''], & \text{if } (M', \phi') \cong (M, \phi) \oplus (M'', \phi'') \\ [M, \phi] = 0, & \text{if } (M, \phi) \text{ is metabolic.} \end{cases}$$

Two forms represent the same Witt class $[M, \phi] = [M', \phi']$ if and only if there exist metabolic forms $(H, \eta)$ and $(H', \eta')$ such that
$$(M \oplus H, \phi \oplus \eta) \cong (M' \oplus H', \phi' \oplus \eta').$$

DEFINITION 3.23. Let $W^\epsilon(R\text{–}A)$ denote the Witt group of the category of representations:
$$W^\epsilon(R\text{–}A) := W^\epsilon((R\text{–}A)\text{-Proj})$$
In particular, $W^\epsilon(\mathbb{Z}\text{–}A) = W^\epsilon(A)$.

REMARK 3.24. If $A$ is a commutative ring then $W^\epsilon(R\text{–}A)$ is a module over $W^1(A)$.

LEMMA 3.25. *An exact duality preserving functor* $(F, \Phi, \eta) : \mathcal{C} \to \mathcal{D}$ *induces a homomorphism of Witt groups:*
$$W^\epsilon(\mathcal{C}) \to W^{\epsilon\eta}(\mathcal{D})$$
$$[M, \phi] \mapsto [F(M), \Phi_M F(\phi)]$$

PROOF. By lemma 3.21, $(F, \Phi, \eta)$ induces an additive functor $H^\epsilon(\mathcal{C}) \to H^{\epsilon\eta}(\mathcal{D})$. We need only prove that if $L$ metabolizes $(M, \phi)$ then $F(L)$ metabolizes $(F(M), \Phi_M F(\phi))$.

Suppose $L = L^\perp$ so there is an exact sequence
$$0 \to L \xrightarrow{j} M \xrightarrow{j^*\phi} L^* \to 0 .$$
Since $F$ is exact the sequence
$$0 \to F(L) \xrightarrow{F(j)} F(M) \xrightarrow{F(j^*\phi)} F(L^*) \to 0$$
is also exact. Observing that
$$\Phi_L F(j^*\phi) = \Phi_L F(j^*) F(\phi) = F(j)^* \Phi_M F(\phi) : F(M) \to F(L)^*,$$
the sequence
$$0 \to F(L) \xrightarrow{F(j)} F(M) \xrightarrow{F(j)\Phi_M F(\phi)} F(L)^* \to 0$$

is exact so $F(L) = F(L)^\perp$. □

For example, the duality preserving functor $(B \otimes_A \_, \Phi, 1) : (R\text{--}A)\text{-Proj} \to (R\text{--}B)\text{-Proj}$ of example 3.20 induces a homomorphism $W^\epsilon(R\text{--}A) \to W^\epsilon(R\text{--}B)$. Setting $R = \mathbb{Z}$, one recovers the homomorphism $W^\epsilon(A) \to W^\epsilon(B)$ defined in section 3.2.

In the Witt group computations of chapter 5 one needs the following two lemmas:

LEMMA 3.26. *If $(M, \phi) \in H^\epsilon(\mathcal{C})$ and $j : L \hookrightarrow M$ is a sublagrangian, i.e.*
$$\phi_L = j^*\phi j = 0 : L \to L^*$$
*then*
$$[M, \phi] = \left[\frac{L^\perp}{L}, \bar{\phi}_{L^\perp}\right] \in W^\epsilon(\mathcal{C})$$
*where $\bar{\phi}_{L^\perp} : \frac{L^\perp}{L} \to \left(\frac{L^\perp}{L}\right)^*$ is induced from the restriction $\phi_{L^\perp} = (j^\perp)^* \phi j^\perp : L^\perp \to (L^\perp)^*$.*

PROOF. The image of the diagonal map $\Delta : L^\perp \to M \oplus L^\perp/L$ is a metabolizer for $(M, -\phi) \oplus (L^\perp/L, \bar{\phi}_{L^\perp})$. □

LEMMA 3.27. *If $(M, \phi) \in H^\epsilon(\mathcal{C})$ and $j : M' \hookrightarrow M$ is a subobject such that the restriction $(M', \phi_{M'})$ is non-singular, i.e. $\phi_{M'} = j^* \phi j : M' \to M'^*$ is an isomorphism, then*
$$(M, \phi) \cong (M', \phi_{M'}) \oplus (M'^\perp, \phi_{M'^\perp}).$$

PROOF. The inclusion $j : M' \hookrightarrow M$ is split by $(j^*\phi j)^{-1} j^*\phi : M \twoheadrightarrow M'$ the kernel of which is $M'^\perp$ so $M = M' \oplus M'^\perp$ as required. □

## 4. Seifert Forms

As we outlined in sections 4.2 and 5 of chapter 1, the $F_\mu$-link cobordism group $C_n(F_\mu)$ is isomorphic to a Witt group of Seifert forms $G^{\epsilon,\mu}(\mathbb{Z})$. We define this Witt group next.

DEFINITION 3.28. Let $\epsilon = +1$ or $-1$ and let $A$ be any ring with involution. A ($\mu$-component) Seifert form over $A$ is a ($\mu+2$)-tuple $(M, \pi_1, \cdots, \pi_\mu, \lambda)$ where $M$ is a finitely generated projective $A$-module, $\pi_1, \cdots, \pi_\mu$ is a set of orthogonal idempotents in $\text{End}_A(M)$ and $\lambda : M \to M^*$ is an $A$-module homomorphism such that $\lambda + \epsilon\lambda^*$ is an isomorphism which commutes with each $\pi_i$. The category of Seifert forms will be denoted $S^\epsilon(A, \mu)$. We usually suppress the component structure writing a Seifert form as $(M, \lambda)$.

A Seifert form $(M, \lambda)$ is said to be metabolic if there exists a metabolizer $L$ for the $\epsilon$-hermitian form $\phi := \lambda + \epsilon\lambda : M \to M^*$ such that
$$\lambda(L)(L) = 0 \quad \text{and} \quad L = L \cap \pi_1 M \oplus \cdots \oplus L \cap \pi_\mu M.$$

LEMMA 3.29. *A summand L of a Seifert form $(M, \lambda)$ is a metabolizer if and only if*
$$L = L^{\perp\lambda} = L^{\perp\phi} \quad \text{and} \quad L = L \cap \pi_1 M \oplus \cdots \oplus L \cap \pi_\mu M.$$
*where $\phi = \lambda + \epsilon\lambda^*$ and $L^{\perp\lambda} = \{x \in M \mid \lambda(x)(y) = \lambda(y)(x) = 0 \ \forall y \in L\}$.*

PROOF. We must show that if $L = L^{\perp\phi}$ then $\lambda(L)(L) = 0$ if and only if $L = L^{\perp\lambda}$. Certainly $L = L^{\perp\lambda}$ implies $\lambda(L)(L) = 0$. Conversely, if we have $\lambda(L)(L) = 0$ then $L \subset L^{\perp\lambda} \subset L^{\perp\phi} = L$. □

DEFINITION 3.30. The Witt group $G^{\epsilon,\mu}(A)$ of ($\mu$-component) Seifert forms is the abelian group generated by isomorphism classes $[M, \lambda]$ of $\mu$-component Seifert forms subject to relations
$$\begin{cases} [M', \lambda'] = [M, \lambda] + [M'', \lambda''], & \text{if } (M', \lambda') \cong (M, \lambda) \oplus (M'', \lambda'') \\ [M, \lambda] = 0, & \text{if } (M, \lambda) \text{ is metabolic.} \end{cases}$$

Recall from section 1.1 that $P_\mu$ denotes the ring
$$P_\mu \cong \mathbb{Z}\left\langle s, \pi_1, \cdots, \pi_\mu \,\Big|\, \sum_{i=1}^\mu \pi_i = 1, \pi_i^2 = \pi_i, \pi_i \pi_j = 0 \text{ for } 1 \leq i, j \leq \mu \right\rangle.$$
which is the path ring of a quiver with $\mu$ vertices and $\mu^2$ arrows. If one endows $P_\mu$ with the involution
$$(21) \qquad\qquad \overline{\pi_i} = \pi_i; \qquad \overline{s} = 1 - s$$
then the Witt group $G^{\epsilon,\mu}(A)$ is in fact isomorphic to the Witt group $W^\epsilon(P_\mu\text{-}A)$ of the representation category:

LEMMA 3.31. *The category $S^\epsilon(A, \mu)$ of Seifert forms is equivalent to the category $H^\epsilon(P_\mu\text{-}A)$ of forms over representations of $P_\mu$. Consequently, there is an isomorphism*
$$\kappa : G^{\epsilon,\mu}(A) \xrightarrow{\cong} W^\epsilon(P_\mu\text{-}A)$$
*which is natural with respect to $A$.*

PROOF. Suppose $(M, \lambda)$ is a $\mu$-component Seifert form. We define a corresponding representation $(M, \rho : P_\mu \to \text{End}_A M)$ and a form $\phi : M \to M^*$ as follows:
$$\phi = \lambda + \epsilon\lambda^*$$
$$\rho(s) = \phi^{-1}\lambda$$
$$\rho(\pi_i^2) = \rho(\pi_i) : M \twoheadrightarrow M_i \rightarrowtail M.$$

By definition $\rho(\pi_i)$ projects $M$ onto its $i$th component $M_i$.

Inversely, one can recover a $\mu$-component Seifert form from a triple $(M, \rho, \phi)$ by setting $\lambda = \phi\rho(s)$. For

$$\begin{aligned}\lambda + \epsilon\lambda^* &= \phi\rho(s) + \epsilon(\phi\rho(s))^* \\ &= (1 - \rho(s)^*)\phi + \rho(s)^*\phi \\ &= \phi.\end{aligned}$$

Metabolic Seifert forms correspond to metabolic $(P_\mu\text{-}A)$-forms which implies the last sentence of the lemma. □

CHAPTER 4

# Morita Equivalence

This expository chapter concerns certain equivalences between categories of modules and hermitian forms.

EXAMPLE 4.1. Suppose $M$ is a simple representation of a ring $R$ over a field $k$. The endomorphism ring $E = \operatorname{End}_{(R\text{--}k)} M$ is a division algebra, and there is a correspondence

$$\overbrace{M \oplus \cdots \oplus M}^{r} = M^{\oplus r} \longleftrightarrow E^{r} \quad (r \in \mathbb{N})$$

between direct sums of copies of $M$ and (right) $E$-modules.

If $R$ and $k$ are endowed with involutions and $M$ is a simple representation which is $\epsilon$-self-dual, there is a Morita equivalence between the category of $\epsilon$-hermitian forms $M^{\oplus r} \to (M^{\oplus r})^*$ and the category of (symmetric or) hermitian forms $E^{\oplus r} \to (E^{\oplus r})^*$.

In chapter 5 the Witt groups $W^\epsilon(R\text{--}k)$ of forms over representations of $R$ will be reduced by a process of devissage to a direct sum of Witt groups of forms over isotypic representations $M^{\oplus r}$. Morita equivalence allows one to pass to the Witt groups of the division algebra of endomorphisms of $M$. This equivalence is a generalization of the trace construction of Milnor (lemma 1.1 in [69]) which played a crucial role in his analysis of isometries of inner product spaces and hence in the computation of high-dimensional knot cobordism.

In the present chapter, we proceed in greater generality.

NOTATION 4.2. Let $\mathcal{C}$ be an additive category [5, pp12-20]. Each object $M$ of $\mathcal{C}$ generates a full subcategory $\mathcal{C}|_M$ of objects isomorphic to a direct summand of some direct sum $M^{\oplus d}$ of copies of $M$.

This restricted category $\mathcal{C}|_M$ is Morita equivalent to the category of projective right modules over the endomorphism ring of $M$:

$$\mathcal{C}|_M \longleftrightarrow \operatorname{Proj-End}_\mathcal{C}(M)$$

A similar result (theorem 4.7 below) applies to the corresponding categories of hermitian forms.

References for this chapter include Curtis and Reiner [19, Vol I,§3D] for the linear theory and Quebbemann Scharlau and Schulte [80, p271], Scharlau [93, §7.4] or Knus [44, §I.9,ch.II] for the hermitian theory.

## 1. Linear Morita Equivalence

DEFINITION 4.3. An additive category $\mathcal{C}$ is *idempotent complete* if each idempotent $p^2 = p : N \to N$ has a splitting $N \xrightarrow{i} N' \xrightarrow{j} N$ such that $ji = p$ and $ij = \text{id}$.

All the categories we shall need, such as the category $(R\text{–}A)$-Proj of representations of $R$ by finitely generated projective $A$-modules, are idempotent complete.

THEOREM 4.4 (Linear Morita Equivalence). *Suppose that $M$ is an object in an additive category $\mathcal{C}$ and $\mathcal{C}|_M$ is idempotent complete. Let $E = \text{End}_{\mathcal{C}} M$. Then $\mathcal{C}|_M$ is equivalent to the category $\text{Proj-}E$ of finitely generated projective right $E$-modules.*

PROOF. $M$ admits an action on the left by $E$ and if $N \in \mathcal{C}$ then $\text{Hom}_{\mathcal{C}}(M, N)$ is a right $E$-module under composition. The functor

$$\text{Hom}_{\mathcal{C}}(M, \_) : \mathcal{C}|_M \to \text{Proj-}E$$

has inverse $\_ \otimes_E M$. $\square$

COROLLARY 4.5. *If $\mathcal{C}$ is an exact category and $\mathcal{C}|_M$ is idempotent complete then*

$$K_0(\mathcal{C}|_M) \cong K_0(E^o).$$

REMARK 4.6 (Naturality of Morita Equivalence). If $F : \mathcal{C} \to \mathcal{D}$ is an additive (covariant) functor then $F$ induces a ring homomorphism $F : \text{End}_{\mathcal{C}} M \to \text{End}_{\mathcal{D}} F(M)$, and the following square commutes up to natural isomorphism:

$$\begin{array}{ccc} \mathcal{C}|_M & \longrightarrow & \mathcal{D}|_{F(M)} \\ \simeq \downarrow & & \downarrow \simeq \\ \text{Proj-}\text{End}_{\mathcal{C}}(M) & \longrightarrow & \text{Proj-}\text{End}_{\mathcal{D}} F(M). \end{array}$$

In particular, there is a commutative square

$$\begin{array}{ccc} K_0(\mathcal{C}|_M) & \longrightarrow & K_0(\mathcal{D}|_{F(M)}) \\ \cong \downarrow & & \downarrow \cong \\ K_0(\text{End}_{\mathcal{C}}(M)^o) & \longrightarrow & K_0(\text{End}_{\mathcal{D}} F(M)^o). \end{array}$$

PROOF. The map

$$\text{Hom}_{\mathcal{C}}(M, N) \otimes_{\text{End}_{\mathcal{C}} M} \text{End}_{\mathcal{D}} F(M) \to \text{Hom}_{\mathcal{D}}(F(M), F(N))$$
$$\phi \otimes f \mapsto F(\phi) f$$

is natural for $N \in \mathcal{C}|_M$ and is an isomorphism because

$$\text{End}_{\mathcal{C}}(M) \otimes_{\text{End}_{\mathcal{C}} M} \text{End}_{\mathcal{D}} F(M) \cong \text{End}_{\mathcal{D}} F(M). \quad \square$$

The functors we shall be considering are those induced by a change of ground ring:
$$F : (R\text{--}A)\text{-Proj} \to (R\text{--}B)\text{-Proj}$$
$$(M, \rho) \mapsto (B \otimes_A M, \rho_B)$$
where $\rho_B(r)(b \otimes x) = b \otimes \rho(r)x$ for all $r \in R$, $b \in B$ and $x \in M$. There is a commutative diagram

$$\begin{array}{ccc}
K_0\left((R\text{--}A)\text{-Proj}|_M\right) & \longrightarrow & K_0\left((R\text{--}B)\text{-Proj}|_{B \otimes M}\right) \\
\cong \updownarrow & & \updownarrow \cong \\
K_0\left((\text{End}_{(R\text{--}A)} M)^o\right) & \longrightarrow & K_0\left((\text{End}_{(R\text{--}B)} B \otimes M)^o\right)
\end{array}$$

In particular, if $A$ and $B$ are commutative and if $B$ is flat as an $A$-module then there is an isomorphism
$$B \otimes_A \text{End}_{(R\text{--}A)} M \cong \text{End}_{(R\text{--}B)}(B \otimes M)$$
$$b \otimes f \mapsto (b' \otimes x \mapsto bb' \otimes f(x)) .$$

## 2. Hermitian Morita equivalence

Suppose $\mathcal{C}$ is a hermitian category and $M \cong M^*$ is a self-dual object in $\mathcal{C}$. The full subcategory $\mathcal{C}|_M$ inherits a hermitian structure.

THEOREM 4.7 (Hermitian Morita Equivalence). *Let $\eta = +1$ or $-1$. Suppose that $(M, b)$ is a non-singular $\eta$-hermitian form in a hermitian category $\mathcal{C}$, and assume further that the hermitian subcategory $\mathcal{C}|_M$ is idempotent complete. Let $E = \text{End}_\mathcal{C} M$ be endowed with the adjoint involution $f \mapsto \overline{f} = b^{-1} f^* b$. Then there is an equivalence of hermitian categories*
$$\Theta_{M,b} = (\text{Hom}(M, \_), \Phi, \eta) : \mathcal{C}|_M \to E\text{-Proj}$$

The precise definition II.6 of equivalence of hermitian categories is given in appendix II.

PROOF OF THEOREM 4.7. Identifying Proj-$E$ with $E$-Proj there is a functor
$$\text{Hom}(M, \_) : \mathcal{C}|_M \to E\text{-Proj}$$
which is an equivalence of categories by theorem 4.4 above. We aim to extend this functor to an equivalence of hermitian categories. By proposition II.7 of appendix II any duality preserving functor $(\text{Hom}(M, \_), \Phi, \eta)$ is automatically an equivalence of hermitian categories.

One can define $\Phi : \text{Hom}(M, \_^*) \to \text{Hom}(M, \_)^*$ by asserting that $\Phi_N$ should be the composite of the following natural isomorphisms:

(22)  $\text{Hom}_\mathcal{C}(M, N^*) \to \text{Hom}_\mathcal{C}(N, M) \to \text{Hom}_E(\text{Hom}_\mathcal{C}(M, N), E)$
$$\gamma \mapsto b^{-1}\gamma^*; \qquad \delta \mapsto (\alpha \mapsto \delta\alpha). \qquad \square$$

COROLLARY 4.8. *If $\mathcal{C}$ is a hermitian admissible subcategory of an abelian category, $(M, b)$ is a non-singular $\eta$-hermitian form and $\mathcal{C}|_M$ is idempotent complete then there is an isomorphism*

$$\Theta_{M,b} : W^\epsilon(\mathcal{C}|_M) \xrightarrow{\cong} W^{\epsilon\eta}(\operatorname{End}_\mathcal{C} M);$$
$$[N, \phi] \mapsto [\operatorname{Hom}(M, N), \Phi_N \phi_*]$$

*where $\phi_* : \operatorname{Hom}(M, N) \to \operatorname{Hom}(M, N^*)$ and $(\Phi_N \phi_*)(\alpha)(\beta) = \epsilon b^{-1} \alpha^* \phi \beta \in \operatorname{End}_\mathcal{C} M$ for all $\alpha, \beta \in \operatorname{Hom}(M, N)$.*

PROOF. The corollary follows from lemma 3.25 because $\Theta_{M,b}$ is an exact functor. □

KNOT THEORY EXAMPLE 4.9. A simple self-dual representation over $\mathbb{Q}$ of $P_1 = \mathbb{Z}[s]$ may be written

$$K = \frac{\mathbb{Q}[s]}{p(s)} \cong \frac{\mathbb{Q}[s]}{p(1-s)} = \operatorname{Hom}_\mathbb{Q}(K, \mathbb{Q})$$

where $p \in \mathbb{Q}[s]$ is an irreducible polynomial.

Milnor's trace construction (lemma 1.1 in [**69**]) is a special case of hermitian Morita equivalence. The trace construction turns a symmetric form $\phi : K^r \to \operatorname{Hom}_\mathbb{Q}(K^r, \mathbb{Q})$ which respect the action of $s$ into a hermitian form over the field $K = \frac{\mathbb{Q}[s]}{p}$ with involution $s \mapsto 1 - s$. The key to the construction is the following isomorphism

$$\operatorname{Hom}_K(K^r, K) \xrightarrow{\cong} \operatorname{Hom}_\mathbb{Q}(K^r, \mathbb{Q})$$
$$\delta \mapsto \operatorname{Trace}_{K/\mathbb{Q}} \circ \delta.$$

This isomorphism is inverse to (22) above if one defines $b : K \to \operatorname{Hom}_\mathbb{Q}(K, \mathbb{Q})$ by $b(x)(y) = \operatorname{Trace}(x.y)$ for all $x, y \in K$.

The endomorphism ring of $K$ is just $K$ itself. Since $K$ is commutative any choice of isomorphism $b : K \to \operatorname{Hom}_\mathbb{Q}(K, \mathbb{Q})$ which respects the representations - i.e. $b(sm)(m') = b(m)((1-s)m')$ - will induce the same adjoint involution

$$K \to K; \quad s \mapsto b^{-1} s^* b = 1 - s.$$

REMARK 4.10 (Naturality). If $(F, \Phi, \eta) : \mathcal{C} \to \mathcal{D}$ is a duality preserving functor and $(M, b) \in H^\eta(\mathcal{C})$ then $(F(M), \Phi_M F(b)) \in H^\eta(\mathcal{D})$ and the following diagram of hermitian Morita equivalences

$$\begin{array}{ccc} \mathcal{C}|_M & \longrightarrow & \mathcal{D}|_{F(M)} \\ \updownarrow & & \updownarrow \\ \operatorname{End}_\mathcal{C}(M)\text{-Proj} & \longrightarrow & \operatorname{End}_\mathcal{D} F(M)\text{-Proj.} \end{array}$$

commutes up to natural isomorphism (of duality preserving functors). If $F$ is exact then the following square also commutes:

$$\begin{array}{ccc} W^\epsilon(\mathcal{C}|_M) & \longrightarrow & W^\epsilon(\mathcal{D}|_{F(M)}) \\ \cong \updownarrow & & \updownarrow \cong \\ W^{\epsilon\eta}(\mathrm{End}_{\mathcal{C}}(M)) & \longrightarrow & W^{\epsilon\eta}(\mathrm{End}_{\mathcal{D}} F(M)). \end{array}$$

In particular, a homomorphism $A \to B$ of rings with involution induces a commutative diagram

$$\begin{array}{ccc} W^\epsilon((R\text{–}A)\text{-Proj}|_M) & \longrightarrow & W^\epsilon((R\text{–}B)\text{-Proj}|_{B\otimes M}) \\ \cong \updownarrow & & \updownarrow \cong \\ W^{\epsilon\eta}(\mathrm{End}_{(R\text{–}A)} M) & \longrightarrow & W^{\epsilon\eta}(\mathrm{End}_{(R\text{–}B)} B \otimes M). \end{array}$$

As we remarked above, if $A \to B$ is a flat homomorphism of commutative rings then $\mathrm{End}_{(R\text{–}B)} B \otimes M$ is naturally isomorphic to $B \otimes_A \mathrm{End}_{(R\text{–}A)} M$.

# CHAPTER 5

# Devissage

In this chapter we assume that $\mathcal{C}$ is an abelian category. Roughly speaking, an abelian category is a category in which every morphism has a kernel, an image and a cokernel. The standard reference for a precise definition is the book of H.Bass [**5**, p21].

We shall assume further that each object in $\mathcal{C}$ has finite length. We discuss the Jordan-Hölder theorem and the corresponding structure theorem in the hermitian setting. The example to keep in mind is the representation category $\mathcal{C} = (R\text{-}k)\text{-Proj}$ where $k$ is a field.

Recall from section 2.2 of chapter 3 that $K_0(\mathcal{C})$ is the abelian group generated by the isomorphism classes of $\mathcal{C}$ with relations corresponding to exact sequences in $\mathcal{C}$.

THEOREM 5.1 (Jordan-Hölder). *Suppose $\mathcal{C}$ is an abelian category with ascending and descending chain conditions. There is a canonical isomorphism*

$$K_0(\mathcal{C}) \cong \bigoplus_{M \in \mathcal{M}^s(\mathcal{C})} K_0(\mathcal{C}|_M)$$

*where $\mathcal{M}^s(\mathcal{C})$ is the set of isomorphism classes of simple objects $M$ in $\mathcal{C}$.*

PROOF. See, for example Lang [**50**, p22,157]. □

Every infinite ordered set contains either an infinite ascending chain $i_1 < i_2 < \cdots$ or an infinite descending chain $i_1 > i_2 > \cdots$. The hypothesis that $\mathcal{C}$ has ascending and descending chain conditions implies that every object $N$ of $\mathcal{C}$ has a finite composition series

$$0 = N_0 \subset N_1 \subset \cdots \subset N_l = N$$

such that each subfactor $N_i/N_{i-1}$ is simple. The content of theorem 5.1 is that these subfactors are uniquely determined up to reordering. The natural number $l$ is also uniquely determined and is called the composition length of $N$.

COROLLARY 5.2. *$K_0(\mathcal{C})$ is isomorphic to a direct sum of copies of $\mathbb{Z}$, one for each simple isomorphism class:*

$$K_0(\mathcal{C}) \cong \mathbb{Z}^{\oplus \mathcal{M}^s(\mathcal{C})} \left( \cong \bigoplus_{M \in \mathcal{M}^s(\mathcal{C})} K_0(\operatorname{End}_{\mathcal{C}}(M)^o) \right).$$

The corresponding theorem for Witt groups is the following:

THEOREM 5.3 (Hermitian Devissage). *Suppose $\mathcal{C}$ is an abelian hermitian category with ascending and descending chain conditions. There is an isomorphism of Witt groups*

$$W^\epsilon(\mathcal{C}) \cong \bigoplus_{M \in \overline{\mathcal{M}}^s(\mathcal{C},\epsilon)} W^\epsilon(\mathcal{C}|_M)$$

*where $\overline{\mathcal{M}}^s(\mathcal{C}, \epsilon)$ denotes the set of isomorphism class of simple $\epsilon$-self-dual objects in $\mathcal{C}$.*

COROLLARY 5.4. *Under the hypotheses of theorem 5.3 The Witt group of $\mathcal{C}$ is isomorphic to a direct sum of Witt groups of division rings:*

$$W^\epsilon(\mathcal{C}) \cong \bigoplus_{M \in \overline{\mathcal{M}}^s(\mathcal{C},\epsilon)} W^1(\mathrm{End}_\mathcal{C} M) .$$

PROOF OF COROLLARY. For each simple $\epsilon$-self-dual object $M$, choose an $\epsilon$-hermitian form $(M, b)$ and apply theorem 4.7. □

When $M$ is simple there is often little distinction between the adjectives self-dual and $\epsilon$-self-dual (e.g. [**93**, p255]):

LEMMA 5.5. *Suppose $M$ is a simple self-dual object in an abelian category $\mathcal{C}$ and $2.\,\mathrm{id} : M \to M$ is an isomorphism. Then $M$ is either $1$-self-dual or $(-1)$-self-dual or both.*

PROOF. Let $f : M \to M^*$ be an isomorphism. Since

$$f = \frac{1}{2}(f + f^*) + \frac{1}{2}(f - f^*)$$

the forms $(f + f^*)$ and $(f - f^*)$ cannot both be zero, so one or other is an isomorphism. □

Suppose next that $M$ is any $\epsilon$-self-dual object. If $2.\,\mathrm{id}_M = 0$ then there is no distinction between $\epsilon$-self-dual and $(-\epsilon)$-self-dual. On the other hand, if $2.\,\mathrm{id}_M \neq 0$ the next lemma gives a criterion for $M$ to be $(-\epsilon)$-self dual. Recall that an $\epsilon$-hermitian form $b : M \to M^*$ induces an involution

$$I_b : \mathrm{End}(M) \to \mathrm{End}(M)$$
$$f \mapsto b^{-1} f^* b.$$

LEMMA 5.6. *An $\epsilon$-self-dual object $M$ fails to be $-\epsilon$-self-dual if and only if some (and therefore every) $\epsilon$-hermitian form $b : M \to M^*$ induces the identity involution $I_b = \mathrm{id}$ on $\mathrm{End}_\mathcal{C} M$.*

PROOF. Suppose $\epsilon b^* = b : M \to M^*$ and $f \neq I_b(f)$ for some endomorphism $f \in \mathrm{End}_\mathcal{C} M$. Let $c = f - I_b(f)$ and observe that $I_b(c) = -c$. The composite $b' = bc$ is a $(-\epsilon)$-hermitian form.

Conversely, suppose $b$ is an $\epsilon$-hermitian form and $b'$ is a $(-\epsilon)$-hermitian form. Setting $f = b^{-1}b'$ we have

$$b^{-1}f^*b = b^{-1}(b')^*b^{-*}b = b^{-1}(-\epsilon)b'\epsilon b^{-1}b = -b^{-1}b' = -f.$$
□

In particular, lemmas 5.5 and 5.6 imply that every self-dual simple complex representation is both 1-self-dual and $(-1)$-self-dual.

COROLLARY 5.7. $W^\epsilon(R\text{–}\mathbb{C}^-)$ *is isomorphic to a direct sum of copies of* $\mathbb{Z}$, *one for each isomorphism class of simple self-dual complex representations of* $R$:

$$W^\epsilon(R\text{–}\mathbb{C}^-) \cong \mathbb{Z}^{\oplus \overline{\mathcal{M}}^s(R\text{–}\mathbb{C})} .$$

PROOF. Set $\mathcal{C} = (R\text{–}\mathbb{C}^-)\text{-Proj}$ and observe that each of the endomorphism rings $\text{End}_\mathcal{C} M$ is isomorphic to $\mathbb{C}^-$. □

We have proved a part of proposition 2.3:

COROLLARY 5.8. $G^{\epsilon,\mu}(\mathbb{C}^-) \cong \mathbb{Z}^{\oplus \infty}$.

A generalization of Pfister's theorem (see chapter 7 below) will imply that the composite

$$C_{2q-1}(F_\mu) \cong G^{\epsilon,\mu}(\mathbb{Z}) \to G^{\epsilon,\mu}(\mathbb{C}^-) \cong \mathbb{Z}^{\oplus \infty}$$

with $q \geq 2$ and $\epsilon = (-1)^q$ defines a *complete* set of signature invariants.

We have also proved a part of proposition 2.6:

COROLLARY 5.9. *There is an isomorphism*

$$G^{\epsilon,\mu}(\mathbb{Q}) \cong \bigoplus_M W^1(\text{End}(M))$$

*with one summand for each isomorphism class of $\epsilon$-self-dual simple rational representation of* $P_\mu$.

To prove theorem 5.3 we need the following lemma

LEMMA 5.10. *Suppose $\mathcal{C}$ is a hermitian abelian category. A stably metabolic $\epsilon$-hermitian form over $\mathcal{C}$ is metabolic. In other words, $(M, \phi)$ represents the zero element of $W^\epsilon(\mathcal{C})$ if and only if $(M, \phi)$ is metabolic.*

PROOF. (e.g. [80, Cor 6.4]) Suppose $L$ metabolizes $(H, \eta)$ and $L'$ metabolizes $(M \oplus H, \phi \oplus \eta)$. Let $L'' = L + L^\perp \cap L' \subset M \oplus H$ and observe that

$$L''^\perp = L^\perp \cap (L^\perp \cap L')^\perp = L^\perp \cap (L + L'^\perp) = L^\perp \cap (L + L')$$
$$= L + L^\perp \cap L' \quad \text{since } L \subset L^\perp$$
$$= L''$$

so $L''$ metabolizes $(M \oplus H, \phi \oplus \eta)$. Since $L''$ contains $L$ we have $L'' = L \oplus (L'' \cap M)$ whence $L'' \cap M$ metabolizes $(M, \phi)$. □

PROOF OF THEOREM 5.3. A general element of $\bigoplus W^\epsilon(\mathcal{C}|_M)$ can be expressed as a formal sum $\sum_{i=1}^r [N_i, \phi_i]$ where each form $(N_i, \phi_i)$ represents an

element of $W^\epsilon(\mathcal{C}|_{M_i})$. Let
$$j : \bigoplus_M W^\epsilon(\mathcal{C}|_M) \to W^\epsilon(\mathcal{C})$$
$$\sum_{i=1}^r [N_i, \phi_i] \mapsto \left[\bigoplus_{i=1}^r N_i, \bigoplus_{i=1}^r \phi_i\right].$$

We check first that $j$ is well-defined and injective. Indeed, if each $(N_i, \phi_i)$ is metabolic with metabolizer $L_i$ say then $L = \bigoplus L_i$ metabolizes $(N, \phi) = (\bigoplus_{i=1}^r N_i, \bigoplus_{i=1}^r \phi_i)$. Conversely, suppose $L$ is a metabolizer for $(N, \phi)$. Since $N$ is semisimple, $L$ is also semisimple. Moreover, each simple subobject of $L$ is isomorphic to one of the $M_i$ and is contained in the corresponding $M_i$-isotypic summand $N_i \subset N$. Thus $L = \bigoplus_i L \cap N_i$ and $L \cap N_i$ metabolizes $(N_i, \phi_i)$.

It remains to show that $j$ is surjective. Let $(N, \phi)$ be a representative of a Witt class in $W^\epsilon(\mathcal{C})$ and let $l = l(N)$ be the composition length of $N$. We argue by induction on $l$. If $l=1$ then $N$ is simple and there is nothing to prove. If $l(N) > 1$, let $j : M_i \hookrightarrow N$ be the inclusion of a simple subobject. The composite
$$M_i \hookrightarrow N \xrightarrow{\phi} N^* \twoheadrightarrow M_i^*$$
is either the zero map or an isomorphism. In the former case
$$[N, \phi] = \left[\frac{M_i^\perp}{M_i}, \overline{\phi}_{M_i^\perp}\right]$$
by lemma 3.26. In the latter case the restriction $(M_i, \phi_{M_i})$ is non-singular so $(W, \phi) \cong (M_i, \phi_{M_i}) \oplus (M_i^\perp, \phi_{M_i^\perp})$ by lemma 3.27. Either way, the inductive hypothesis applies. $\square$

REMARK 5.11 (Naturality). If $\mathcal{C}$ and $\mathcal{D}$ satisfy the hypotheses of theorem 5.1 and $F : \mathcal{C} \to \mathcal{D}$ is an exact functor, then there is a commutative diagram
$$\begin{array}{ccc} K_0(\mathcal{C}) & \longrightarrow & K_0(\mathcal{D}) \\ \cong \updownarrow & & \uparrow \\ \bigoplus_M K_0(\mathcal{C}|_M) & \longrightarrow & \bigoplus_M K_0(\mathcal{D}|_{F(M)}) \end{array}$$
with both direct sums indexed over the simple isomorphism classes in $\mathcal{C}$.

Analogously, if $\mathcal{C}$ and $\mathcal{D}$ satisfy the hypotheses of theorem 5.3 and $(F, \Phi, \eta) : \mathcal{C} \to \mathcal{D}$ is an exact duality preserving functor then there is a commutative diagram
$$\begin{array}{ccc} W^\epsilon(\mathcal{C}) & \longrightarrow & W^{\epsilon\eta}(\mathcal{D}) \\ \cong \updownarrow & & \uparrow \\ \bigoplus_M W^\epsilon(\mathcal{C}|_M) & \longrightarrow & \bigoplus_M W^{\epsilon\eta}(\mathcal{D}|_{F(M)}). \end{array}$$

# CHAPTER 6

# Varieties of Representations

The aim of this chapter is to give some geometric structure to the set of signature invariants of Seifert forms over $\mathbb{C}^-$. In chapter 9 characters will be used to identify the 'algebraically integral' representations relevant to Seifert forms over $\mathbb{Z}$ and hence to boundary link cobordism.

NOTATION 6.1. Let $\mathcal{M}(R) = \mathcal{M}(R\text{–}\mathbb{C})$ denote the set of isomorphism classes of semisimple representations of a ring $R$ over $\mathbb{C}$. We denote the subset of self-dual representations $\overline{\mathcal{M}}(R) \subset \mathcal{M}(R)$. Let $\mathcal{M}^s(R) \subset \mathcal{M}(R)$ be the subset of simple representations and let $\overline{\mathcal{M}}^s(R) \subset \overline{\mathcal{M}}(R)$ denote the simple self-dual representations.

In this chapter we focus on the case $R = P_\mu$.

KNOT THEORY EXAMPLE 6.2. When $\mu = 1$ we have $P_1 = \mathbb{Z}[s]$ and there is a correspondence

$$\mathbb{C} \leftrightarrow \mathcal{M}^s(\mathbb{Z}[s])$$
$$\nu \leftrightarrow \mathbb{C}[s]/(s-\nu)$$

which parameterizes the isomorphism classes of simple representations and identifies $\mathcal{M}^s(\mathbb{Z}[s])$ with one-dimensional affine space.

The self-dual representations $\overline{\mathcal{M}}(\mathbb{Z}[s])$ correspond to the points $\nu \in \mathbb{C}$ which satisfy the equation $\bar{\nu} = 1-\nu$. This equation defines a one-dimensional real variety

$$\{1/2 + ib \mid b \in \mathbb{R}\} \subset \mathbb{C}$$

(compare Ranicki [**84**, p609]). We recall the notion of a real variety in definition 6.10 below.

The aim here is to give some analogous algebraic structure to $\mathcal{M}^s(P_\mu)$ and $\overline{\mathcal{M}}^s(P_\mu)$, when $\mu \geq 2$, in the framework of Mumford's geometric invariant theory [**72**]. In chapter 10, explicit computations are given for two particular dimension vectors $\alpha$.

Since $P_\mu$ is the path ring of a quiver (see lemma 3.4) a description of $\mathcal{M}^s(P_\mu)$ can be read off from the work of Le Bruyn and Procesi [**51**] (see also Procesi [**78**]) who applied the Étale slice machinery of D. Luna ([**66**], [**67**]) to study the semisimple representations of quivers.

Given a quiver $Q$ and a fixed dimension vector $\alpha : Q_0 \to \mathbb{Z}_{\geq 0}$ the isomorphism classes of semisimple representations of $Q$ with dimension vector $\alpha$ correspond to the points in an affine algebraic variety (which is not smooth in general). The semisimple isomorphism classes $\mathcal{M}(P_\mu)$ are therefore a disjoint union of affine varieties $\mathcal{M}(P_\mu, \alpha)$, one for each dimension vector $\alpha$.

Each variety 'admits a finite stratification into locally closed smooth irreducible subvarieties corresponding to the different types of semisimple decompositions of dimension vector $\alpha$. Moreover, one stratum lies in the closure of another if the corresponding representations are deformations' ([**51**, p586]). In particular, if there exist simple representations with dimension vector $\alpha$ then they form a dense (Zariski) open smooth subvariety of the variety of semisimple representations.

A duality functor on $(Q\text{-}\mathbb{C}^-)$-Proj induces an involution on each variety of semisimple representations. The subset invariant under the involution is then a real algebraic variety.

## 1. Existence of Simple Representations

The first question is whether any self-dual simple representations exist with dimension vector $\alpha$. When $\mu \geq 2$ the answer is usually yes:

LEMMA 6.3. *Let* $\alpha : \{x_1, \cdots, x_\mu\} \to \mathbb{Z}_{\geq 0}$; $x_i \mapsto \alpha_i$ *be a dimension vector for the quiver* $P_\mu$. *The following are equivalent:*

(1) $\overline{\mathcal{M}^s}(P_\mu, \alpha) \neq \emptyset$.
(2) $\mathcal{M}^s(P_\mu, \alpha) \neq \emptyset$.
(3) *Either* $|\text{Support}(\alpha)| \geq 2$ *or* $\alpha = \delta^i$ *for some* $i$, *where* $\delta^i_j = 1$ *if* $j = i$ *and* $\delta^i_j = 0$ *if* $j \neq i$.

PROOF. $1 \Rightarrow 2$: Immediate.
$2 \Rightarrow 3$: If $\text{Support}(\alpha) = \{i\}$ then a representation of $P_\mu$ of dimension vector $\alpha$ is essentially a representation of a polynomial ring $\mathbb{C}[s_{ii}]$ which by the fundamental theorem of algebra is not simple unless $\alpha_i = 1$.
$3 \Rightarrow 1$: The case where $\text{Support}(\alpha) = \{i\}$ is easy so assume $|\text{Support}(\alpha)| \geq 2$. We shall construct a simple self-dual representation $(M, \rho)$ with dimension vector $\alpha$.

Recall that $s$ is the sum of the paths of length one in the quiver $P_\mu$. We denote by $s_{ij}$ the arrow from vertex $x_j$ to vertex $x_i$. For each $i \in \text{Support}(\alpha)$ let

$$\rho(s_{ii}) = \begin{pmatrix} \nu_1 & 0 & \cdots & 0 \\ 0 & \nu_2 & \cdots & 0 \\ \vdots & \vdots & \ddots & \vdots \\ 0 & 0 & \cdots & \nu_{\alpha_i} \end{pmatrix}$$

where $\nu_1, \cdots, \nu_{\alpha_i}$ are distinct elements of the set $\{1/2 + bi \mid b \in \mathbb{R}\}$. When $i, j \in \text{Support}(\alpha)$ and $i \neq j$, let

$$\rho(s_{ij}) = \text{sign}(i-j) \begin{pmatrix} 1 & 1 & \cdots & 1 \\ 1 & 1 & \cdots & 1 \\ \vdots & \vdots & \ddots & \vdots \\ 1 & 1 & \cdots & 1 \end{pmatrix}.$$

The total matrix $\rho(s)_{ij} = \rho(s_{ij})$ satisfies $\rho(s) = 1 - \overline{\rho(s)}^t$ and so $(M, \rho)$ is self-dual.

We must check that $M$ is simple. Regarding $\pi_i M$ as a representation of a polynomial ring $\mathbb{Z}[s_{ii}]$ for each $i \in \text{Support}(\alpha)$, we have

$$\pi_i M \cong \frac{\mathbb{C}[s_{ii}]}{(s_{ii} - \nu_1)} \oplus \cdots \oplus \frac{\mathbb{C}[s_{ii}]}{(s_{ii} - \nu_{\alpha_i})},$$

a direct sum of simple representations of $\mathbb{Z}[s_{ii}]$ no two of which are isomorphic. Given a subrepresentation $M' \subset M$, it suffices to show that $\pi_i M' = 0$ or $\pi_i M$ for each $i$.

Suppose that $\pi_i M' \neq 0$. Since $|\text{Support}(\alpha)| \geq 2$ we can choose an element $j \in \text{Support}(\alpha)$ with $j \neq i$ and we find

$$\langle (1, 1, \cdots, 1) \rangle \in \rho(s_{ij})\rho(s_{ji})(\pi_i M') \subset \pi_i M'$$

which implies that $\pi_i M' = \pi_i M$ as required. □

## 2. Semisimple Representations of Quivers

Let us fix a quiver $Q$, a dimension vector $\alpha : Q_0 \to \mathbb{Z}_{\geq 0}$ and a family of vector spaces $\{M_x = \mathbb{C}^{\alpha_x}\}_{x \in Q_0}$. As usual, let $m = \sum_{x \in Q_0} \alpha_x$. Dimension vector $\alpha$ representations of $Q$ correspond to families of linear maps

$$\{f_e\}_{e \in Q_1} \in R(Q, \alpha) = \bigoplus_{e \in Q_1} \text{Hom}(\mathbb{C}^{\alpha_{h(e)}}, \mathbb{C}^{\alpha_{t(e)}}),$$

points in an affine space of dimension $\sum_{e \in Q_1} \alpha_{h(e)} \alpha_{t(e)}$. The coordinate ring of this affine space is a polynomial ring $\mathbb{C}[X]$, where $X$ is a set of (commuting) indeterminates.

Isomorphism classes of representations correspond to orbits for the action of the reductive group

$$\text{GL}(\alpha) = \prod_{i=1}^{\mu} \text{GL}(\alpha_i)$$

by conjugation on $R(Q, \alpha)$, i.e. $g.f_e = g_{t(e)} f_e g_{h(e)}^{-1}$ where $e \in Q_1$ and $g = \{g_x\}_{x \in Q_0}$. By Mumford's theory the closed orbits correspond to semisimple isomorphism classes.

The ring of invariants $\mathbb{C}[X]^{\text{GL}(\alpha)}$ is the coordinate ring of a variety

$$\mathcal{M}(Q, \alpha) = R(Q, \alpha) // \text{GL}(\alpha)$$

which parameterizes these semisimple isomorphism classes; the inclusion of $\mathbb{C}[X]^{\text{GL}(\alpha)}$ in $\mathbb{C}[X]$ induces a categorical quotient $R(Q, \alpha) \twoheadrightarrow \mathcal{M}(Q, \alpha)$, the universal $\text{GL}(\alpha)$-invariant morphism. The Zariski topology on $\mathcal{M}(Q, \alpha)$ coincides with the quotient topology so $\mathcal{M}(Q, \alpha)$ is irreducible.

THEOREM 6.4 (Procesi and Le Bruyn). *$\mathbb{C}[X]^{\text{GL}(\alpha)}$ is generated by traces of oriented cycles in the quiver of length at most $m^2 = (\sum \alpha_i)^2$.*

PROOF. See [**51**, Theorem 1]. □

It follows from further work of Procesi [**79**] that all the relations between these traces can be deduced from Cayley-Hamilton relations of $m \times m$ matrices.

As we remarked above, there is a Luna stratification of $\mathcal{M}(Q, \alpha)$ into smooth locally closed subvarieties, one for each representation type:

DEFINITION 6.5. A semisimple representation
$$M = M_1^{\oplus d_1} \oplus \cdots \oplus M_k^{\oplus d_k}$$
is said to be of *representation type* $\tau = (d_1, \beta_1; \cdots ; d_k, \beta_k)$ if each $M_i$ is simple and $\beta_i$ is the dimension vector of $M_i$.

The stratum of representations of type $\tau'$ is contained in the closure of the type $\tau$ stratum if and only if $\tau'$ can be obtained from $\tau$ by a sequence of deformations each decomposing some vector $\beta_i$ as a sum of smaller dimension vectors $\beta_i = \beta_i^{(1)} + \beta_i^{(2)}$ (and perhaps regrouping if $\beta_i^{(j)} = \beta_k$ for some $j$ and $k$). In particular, the stratum of simple isomorphism classes, if it exists, contains all other strata in its closure; it is a dense open subset of $\mathcal{M}(Q, \alpha)$.

REMARK 6.6. The dimension of $\mathcal{M}(Q, \alpha)$ is easy to calculate. We are concerned here with those $\alpha$ for which the generic representation $M$ is simple so that $\dim(\operatorname{Aut}_{P_\mu}(M)) = \dim(\operatorname{End}_{P_\mu}(M)) = 1$.

By the orbit-stabilizer theorem
$$\dim\left(\mathcal{M}(Q, \alpha)\right) = \dim\left(R(Q, \alpha)\right) - \dim\left(\operatorname{GL}(\alpha)\right) + \dim(\operatorname{GL}(\alpha)_M)$$
(where $\operatorname{GL}(\alpha)_M$ is the stabilizer of $M$)
$$= \sum_{e \in Q_1} \alpha_{h(e)} \alpha_{t(e)} - \sum_{x \in Q_0} \alpha_x^2 + 1.$$

In particular when $Q = P_\mu$,

$$(23) \quad \dim\left(\mathcal{M}(P_\mu, \alpha)\right) = \sum_{i,j=1}^{\mu} \alpha_i \alpha_j - \sum_{i=1}^{\mu} \alpha_i^2 + 1 = 1 + \sum_{1 \le i < j \le \mu} 2\alpha_i \alpha_j$$

## 3. Self-Dual Representations

We proceed to give the self-dual isomorphism classes $\overline{\mathcal{M}}(P_\mu, \alpha)$ a real algebraic variety structure. Let us first recall some of the basic theory of real algebraic geometry; for further details the reader is referred to Bochnak, Coste and Roy [**8**] or Lam [**48**].

### 3.1. Real Algebraic Sets.
Let $\mathbb{R}[X] = \mathbb{R}[X_1, \cdots, X_m]$ denote a commutative polynomial ring over the real numbers. If $U \triangleleft \mathbb{R}[X]$ then one can define
$$V(U) = \bigcap_{f \in U} \{x \in \mathbb{C}^m \mid f(x) = 0\}.$$
The intersection $V_\mathbb{R}(U) = V(U) \cap \mathbb{R}^m$ is called a real algebraic set; these generate the closed sets in the Zariski topology for $\mathbb{R}^m$. In fact the points

in $V_\mathbb{R}(U)$ correspond to the maximal ideals of $\mathbb{R}[X]$ which both contain $U$ and are real in the following (algebraic) sense (lemma 6.19 below):

DEFINITION 6.7. A commutative ring $A$ is *real* (also called *formally real*) if it has the property
$$a_1^2 + \cdots + a_l^2 = 0 \Rightarrow a_1 = \cdots = a_l = 0$$
for all $a_1, \cdots, a_l \in A$. An ideal $U$ of a commutative ring $A$ is said to be real if and only if the quotient ring $A/U$ is real.

On the other hand, if one is given a subset $S$ of $\mathbb{R}^m$ one can define
$$U(S) = \bigcap_{x \in S} \{f \in \mathbb{R}[X] \mid f(x) = 0\}.$$

To explain the extent to which $V_\mathbb{R}$ and $U$ are mutually inverse, one must introduce the real radical:

DEFINITION 6.8. The *real radical* $\sqrt[r]{U}$ of an ideal $U \triangleleft A$ is by definition
$$\{a \in A \mid a^{2m} + a_1^2 + \cdots + a_l^2 \in U \text{ for some } m \geq 0 \text{ and } a_1, \cdots, a_n \in A\}.$$

In fact $\sqrt[r]{U}$ is the intersection of the real ideals containing $U$.

THEOREM 6.9 (Dubois-Risler Real Nullstellensatz).
*i)* If $U \triangleleft \mathbb{R}[X]$ then $U(V_\mathbb{R}(U)) = \sqrt[r]{U}$.
*ii)* If $S \subset \mathbb{R}^m$ then $V_R(U(S)) = \overline{S}$ where $\overline{S}$ denotes the closure of $S$ for the Zariski topology on $\mathbb{R}^m$.

The dimension of a real algebraic set $V_\mathbb{R}(U)$ is by definition the (Krull) dimension of $\mathbb{R}[X]/\sqrt[r]{U}$. Dubois and Efroymson [21] proved that given a real variety $V_\mathbb{R}(\mathfrak{p})$ one can find a chain of real varieties
$$V_\mathbb{R}(\mathfrak{p}) = V_\mathbb{R}(\mathfrak{p}_d) \supset V_\mathbb{R}(\mathfrak{p}_{d-1}) \supset \cdots \supset V_\mathbb{R}(\mathfrak{p}_0)$$
where $V_\mathbb{R}(\mathfrak{p}_i)$ has dimension $i$; this result confirms that the algebraic definition of dimension is reflected in the real geometry.

The decomposition theory of semi-algebraic sets provides further reassurance. A semi-algebraic set is a subset of $\mathbb{R}^m$ defined by polynomial inequalities in addition to, or in place of, polynomial equations. Any semi-algebraic set decomposes [8, Thm2.3.6, §2.8] as a disjoint union of semi-algebraic sets each semi-algebraically homeomorphic to an open hypercube $(0,1)^d$. The largest occurring value of $d$ coincides with the Krull dimension defined above.

## 3.2. Real Varieties.

DEFINITION 6.10. An (affine) real algebraic variety $V$ is a subset of $\mathbb{R}^m$ of the form $V = V_\mathbb{R}(\mathfrak{p})$ where $\mathfrak{p}$ is a real prime ideal of $\mathbb{R}[X]$.

Defined thus, a real variety is not only irreducible in the Zariski topology, but absolutely irreducible (i.e. $V(\mathfrak{p})$ is also irreducible). Moreover $V_\mathbb{R}(\mathfrak{p})$ is Zariski dense in $V(\mathfrak{p})$.

## 3.3. Smoothness.

A point $\nu$ in an irreducible real algebraic set $V = V_{\mathbb{R}}(U)$ is said to be *regular* if the localized ring

$$T = \left(\frac{\mathbb{R}[X]}{\sqrt[r]{U}}\right)_{(X_1-\nu_1,\cdots,X_n-\nu_n)}$$

is regular. The tangent space to $V$ at $\nu$

$$\bigcap_{f \in U} \left\{ x \in \mathbb{R}^m \,\bigg|\, \sum_{i=1}^m \frac{\partial f}{\partial X_i}(\nu) x_i = 0 \right\}$$

coincides in dimension with $V$ if and only if $\nu$ is regular [8, Prop3.3.6] so a regular point is also described as *non-singular* or *smooth*.

Smooth points give a useful test for reality:

PROPOSITION 6.11. *(e.g. Lam [48, p797]) Let $\mathfrak{p}$ be a prime ideal in $\mathbb{R}[X]$. Then $\mathfrak{p}$ is real if and only if $V_{\mathbb{R}}(\mathfrak{p})$ contains a smooth point.*

## 3.4. Varieties with Involution.

Returning to representations of $P_\mu$, the duality functor $(\_)^*$ on $(P_\mu\text{-}\mathbb{C}^-)$-Proj induces an involution on $\mathcal{M}(P_\mu)$. To be more explicit, the involution (21) on $P_\mu$ induces an involution $I(\theta) = 1 - \bar{\theta}^*$ on each component $R(P_\mu, \alpha)$, and hence induces an involution on the coordinate ring $\mathbb{C}[X]$. Here, $X = \{X_{ij}\}_{1 \le i,j \le n}$ denotes a set of $m^2$ commuting indeterminates and the induced involution

$$I(X_{ij}) = \begin{cases} -X_{ji} & \text{if } j \ne i \\ 1 - X_{ii} & \text{if } j = i \end{cases}$$

may be briefly written $I(f)(X) = \overline{f}(1 - X^t)$.

Let us combine the action of $\mathrm{GL}(\alpha)$ on $\mathbb{C}[X]$ with the involution.

DEFINITION 6.12. *Let $\mathrm{GL}(\alpha) \rtimes \frac{\mathbb{Z}}{2\mathbb{Z}}$ denote the semidirect product, where $\frac{\mathbb{Z}}{2\mathbb{Z}}$ acts on $\mathrm{GL}(\alpha)$ by the formula $I(g) = (g^{-1})^t$.*

LEMMA 6.13. *There is an action of $\mathrm{GL}(\alpha) \rtimes \frac{\mathbb{Z}}{2\mathbb{Z}}$ on $\mathbb{C}[X]$ which extends both the conjugation action of $\mathrm{GL}(\alpha)$ and the involution.*

PROOF. If $f \in \mathbb{C}[X]$ and $g \in \mathrm{GL}(\alpha)$ then

$$(g.I(f))(X) = I(f)(g^{-1}Xg) = \overline{f}(1 - (\bar{g}^{-1}X\bar{g})^t)$$
$$= (\bar{g}^{-1})^t.\overline{f}(1 - X^t) = I((\bar{g}^{-1})^t.f)(X). \quad \square$$

The following lemma says that $I$ induces an involution on $\mathcal{M}(P_\mu, \alpha) = R(P_\mu, \alpha) /\!/ \mathrm{GL}(\alpha)$:

LEMMA 6.14. *The involution $I$ acts on the invariant ring $\mathbb{C}[X]^{\mathrm{GL}(\alpha)}$.*

PROOF. If $f \in \mathbb{C}[X]^{\mathrm{GL}(\alpha)}$ and $g \in \mathrm{GL}(\alpha)$ then

$$(g.I(f))(X) = I((\bar{g}^{-1})^t.f)(X) = I(f)(X). \quad \square$$

## 3. SELF-DUAL REPRESENTATIONS

The orbits of the action of $I$ correspond to the maximal ideals in the invariant ring
$$\left(\mathbb{C}[X]^{\mathrm{GL}(\alpha)}\right)^{\frac{\mathbb{Z}}{2\mathbb{Z}}} = \mathbb{C}[X]^{\mathrm{GL}(\alpha) \times \frac{\mathbb{Z}}{2\mathbb{Z}}};$$
we are interested in the orbits which contain only one element, for they correspond to self-dual isomorphism classes.

Let us now assume that there are simple representations of $P_\mu$ of dimension vector $\alpha$ (see lemma 6.3).

PROPOSITION 6.15. *The isomorphism classes of self-dual semisimple representations of $P_\mu$ with dimension vector $\alpha$ correspond to the real maximal ideals of the invariant ring $\mathbb{C}[X]^{\mathrm{GL}(\alpha) \times \frac{\mathbb{Z}}{2\mathbb{Z}}}$ and hence to the points in a real algebraic variety of dimension*
$$1 + \sum_{1 \leq i < j \leq \mu} 2\alpha_i \alpha_j \ .$$

PROPOSITION 6.16. *The subset $\overline{\mathcal{M}}^s(P_\mu, \alpha)$ of self-dual simple isomorphism classes is a smooth Zariski open subvariety of $\overline{\mathcal{M}}(P_\mu, \alpha)$.*

The remainder of this chapter concerns the proof of propositions 6.15 and 6.16.

Let $A$ denote any finitely generated commutative $\mathbb{C}$-algebra with an involution $I$ which restricts to complex conjugation on $\mathbb{C}$. Let $A_0 = A^{\frac{\mathbb{Z}}{2\mathbb{Z}}} \subset A$ be the involution-invariant $\mathbb{R}$-algebra.

LEMMA 6.17. *There is a natural isomorphism $A \cong \mathbb{C}^- \otimes_\mathbb{R} A_0$ of rings with involution. In particular, $A$ is an integral extension of $A_0$ and $A_0$ is a finitely generated $\mathbb{R}$-algebra.*

PROOF. The natural map $\mathbb{C}^- \otimes_\mathbb{R} A_0 \to A;\ \nu \otimes a \mapsto \nu a$ has inverse
$$A \to \mathbb{C} \otimes_\mathbb{R} A_0; \quad a \mapsto 1 \otimes \frac{1}{2}((a + I(a)) - i \otimes i(a - I(a)). \qquad \square$$

Recall that an ideal $\mathfrak{m}$ of $A$ is said to *lie over* an ideal $\mathfrak{m}_0$ of $A_0$ if and only if $\mathfrak{m} \cap A_0 = \mathfrak{m}_0$.

LEMMA 6.18. *Over each maximal ideal $\mathfrak{m}_0$ of $A_0$ lies either one maximal ideal $\mathfrak{m} \triangleleft A$ with $\mathfrak{m} = I(\mathfrak{m})$ or precisely two distinct maximal ideals $\mathfrak{m}$ and $I(\mathfrak{m})$.*

PROOF. There exists a maximal ideal $\mathfrak{m} \triangleleft A$ with $\mathfrak{m} \cap A_0 = \mathfrak{m}_0$ by the 'lying over' theorem for integral extensions. Although $I(\mathfrak{m})$ may or may not be distinct from $\mathfrak{m}$, $I(\mathfrak{m})$ certainly lies over $\mathfrak{m}_0$. It remains to show that if $\mathfrak{m}' \triangleleft A$ is maximal and $\mathfrak{m}' \cap A_0 = \mathfrak{m}_0$ then $\mathfrak{m}'$ coincides either with $\mathfrak{m}$ or with $I(\mathfrak{m})$. If we assume the contrary then by the Chinese remainder theorem there exists $a \in A$ such that
$$a \equiv 0 \pmod{\mathfrak{m}}, \quad a \equiv 1 \pmod{\mathfrak{m}'} \quad \text{and} \quad a \equiv 1 \pmod{I(\mathfrak{m}')}.$$
The condition $a \in \mathfrak{m}$ implies that $aI(a) \in \mathfrak{m} \cap A_0 = \mathfrak{m}_0$ whereas the other conditions $a \equiv I(a) \equiv 1 \pmod{\mathfrak{m}'}$ have the contradictory implication $aI(a) \notin \mathfrak{m}'$. $\qquad \square$

LEMMA 6.19. *There is precisely one maximal ideal $\mathfrak{m}$ over $\mathfrak{m}_0$ if and only if $\mathfrak{m}_0$ is real.*

PROOF. By lemma 6.17 we may write $A_0 = \frac{\mathbb{R}[Y]}{U}$ and $A = \frac{\mathbb{C}[Y]}{\mathbb{C} \otimes U}$ for some finite set $Y = \{Y_1, \cdots, Y_m\}$ of commuting indeterminates and some ideal $\mathfrak{p}$. Now $\mathfrak{m}$ has the form $(Y_1 - \nu_1, \cdots, Y_m - \nu_m)$ for some $\nu \in \mathbb{C}^m$ and $\mathfrak{m}_0 = I(\mathfrak{m})$ if and only if $\nu \in \mathbb{R}^m$. If $\nu \in \mathbb{R}^m$ then $\mathbb{R}[Y]/(\mathfrak{m} \cap A_0) \cong \mathbb{R}$ is real. On the other hand, if some $\nu_i \notin \mathbb{R}$ then
$$Y_i^2 - (\nu_i + \bar{\nu}_i)Y_i + \nu_i \bar{\nu}_i = (Y_i - \nu_i)(Y_i - \bar{\nu}_i) \in \mathfrak{m} \cap A_0$$
so in $A_0/(\mathfrak{m} \cap A_0)$ we have $(Y_i - \frac{1}{2}(\nu_i + \bar{\nu}_i))^2 = \frac{1}{4}(\nu_i - \bar{\nu}_i)^2 < 0$. □

PROOF OF PROPOSITION 6.15. To obtain a correspondence between self-dual isomorphism classes of representations and real maximal ideals in the invariant ring $\mathbb{C}[X]^{\mathrm{GL}(\alpha) \times \frac{\mathbb{Z}}{2\mathbb{Z}}}$, we apply lemmas 6.18 and 6.19, putting $A = \mathbb{C}[X]^{\mathrm{GL}(\alpha)}$. We must show that $\overline{\mathcal{M}}(P_\mu, \alpha)$ is a real algebraic variety and compute its dimension.

Now $A = \mathbb{C}[X]^{\mathrm{GL}(\alpha)}$ is an integral domain so the invariant ring $A_0 = \mathbb{C}[X]^{\mathrm{GL}(\alpha) \times \frac{\mathbb{Z}}{2\mathbb{Z}}}$ is of the form $\frac{\mathbb{R}[Y]}{\mathfrak{p}}$ where $\mathfrak{p}$ is a prime ideal. We must check that $\mathfrak{p}$ is real; by proposition 6.11 it suffices to prove that $V_\mathbb{R}(\mathfrak{p})$ contains a smooth point.

Lemma 6.3 above assures us that there are self-dual representations among the simple dimension vector $\alpha$ representations, so $\mathbb{C}[X]^{\mathrm{GL}(\alpha)}$ contains a maximal ideal $\mathfrak{m}$ which is both smooth and involution invariant. The proof of lemma 6.17 implies that $\mathfrak{m} = \mathbb{C} \otimes \mathfrak{m}_0$ where $\mathfrak{m}_0 = \mathfrak{m} \cap \mathbb{C}[X]^{\mathrm{GL}(\alpha) \times \frac{\mathbb{Z}}{2\mathbb{Z}}}$ so

$$\dim_\mathbb{R} \left( \frac{\mathfrak{m}_0}{\mathfrak{m}_0^2} \right) = \dim_\mathbb{C} \left( \frac{\mathfrak{m}}{\mathfrak{m}^2} \right) = \text{Krull-dim} \left( \mathbb{C}[X]^{\mathrm{GL}(\alpha)} \right)$$
$$= \text{Krull-dim} \left( \mathbb{C} \otimes_\mathbb{R} \mathbb{C}[X]^{\mathrm{GL}(\alpha) \times \frac{\mathbb{Z}}{2\mathbb{Z}}} \right) = \text{Krull-dim} \left( \mathbb{C}[X]^{\mathrm{GL}(\alpha) \times \frac{\mathbb{Z}}{2\mathbb{Z}}} \right).$$

Thus $\mathfrak{m}_0$ is a smooth real point and so $\mathfrak{p}$ is a real prime.

The (real) dimension of $\overline{\mathcal{M}}(P_\mu, \alpha)$ is equal to the (complex) dimension of $\mathcal{M}(P_\mu, \alpha)$ which, by remark 6.6 is $1 + \sum_{1 \le i < j \le \mu} 2\alpha_i \alpha_j$. □

PROOF OF PROPOSITION 6.16. Note first that the involution $I$ respects the Luna stratification of $\mathcal{M}(P_\mu, \alpha)$; in other words $I$ preserves representation type. The inclusion of $\mathbb{C}[X]^{\mathrm{GL}(\alpha) \times \frac{\mathbb{Z}}{2\mathbb{Z}}}$ in $\mathbb{C}[X]^{\mathrm{GL}(\alpha)}$ induces a quotient

$$q : \mathcal{M}(P_\mu, \alpha) \to \mathcal{M}(P_\mu, \alpha) // \frac{\mathbb{Z}}{2\mathbb{Z}}$$

and the Zariski topology on $\overline{\mathcal{M}}(P_\mu, \alpha)$ coincides with the subspace topology induced by the inclusion of $\overline{\mathcal{M}}(P_\mu, \alpha)$ in $\mathcal{M}(P_\mu, \alpha) // \frac{\mathbb{Z}}{2\mathbb{Z}}$. The image under $q$ of the Zariski open set $\mathcal{M}^s(P_\mu, \alpha)$ is open in $\mathcal{M}(P_\mu, \alpha) // \frac{\mathbb{Z}}{2\mathbb{Z}}$ and its intersection $\overline{\mathcal{M}}^s(P_\mu, \alpha) = \overline{\mathcal{M}}(P_\mu, \alpha) \cap q(\mathcal{M}^s(P_\mu, \alpha))$ is therefore open in $\overline{\mathcal{M}}(P_\mu, \alpha)$. Smoothness of $\overline{\mathcal{M}}^s(P_\mu, \alpha)$ follows (as in the proof of proposition 6.15) from the smoothness of $\mathcal{M}^s(P_\mu, \alpha)$. □

CHAPTER 7

# Generalizing Pfister's Theorem

This chapter initiates discussion of rationality questions proving that the varieties of signatures defined in the preceding chapters are a *complete* set of torsion-free invariants for boundary link cobordism.

To recap, in section 5.2 of chapter 1 and section 4 of chapter 3 we identified the odd-dimensional $F_\mu$-link cobordism group $C_{2q-1}(F_\mu)$ with the Witt group $W^\epsilon(P_\mu\text{-}\mathbb{Z})$ where $\epsilon = (-1)^q$ and chapters 4 and 5 demonstrated an isomorphism $W^\epsilon(R\text{-}\mathbb{C}^-) \cong \mathbb{Z}^{\oplus \overline{\mathcal{M}}^s(R)}$ for any ring $R$ with involution. We will see in lemma 11.1 that $W^\epsilon(R\text{-}\mathbb{Z})$ injects into $W^\epsilon(R\text{-}\mathbb{Q})$. In the present chapter we prove the following:

THEOREM 7.1. *Every element $v$ in the kernel of the natural map*

$$(24) \qquad W^\epsilon(R\text{-}\mathbb{Q}) \to W^\epsilon(R\text{-}\mathbb{C}^-) \cong \mathbb{Z}^{\oplus \overline{\mathcal{M}}^s(R\text{-}\mathbb{C})}$$

*satisfies $2^n v = 0$ for some integer $n$.*

Using number theory one can prove that the kernel of (24) is in fact 8-torsion - see corollary 11.6 below.

In fact we shall prove a generalization of theorem 7.1 substituting for $\mathbb{Q}$ an arbitrary field $k$ (with trivial involution). This generalization is (Morita) equivalent to a result of Scharlau [92, §5] on Witt groups of semisimple algebras. Indeed, our proof follows Scharlau's closely employing his Frobenius reciprocity method. The special case $R = \mathbb{Z}$ is Pfister's theorem (see Scharlau [93, Thm2.7.3 p56]):

THEOREM 7.2 (Pfister). *A symmetric form $\phi$ over a field $k$ represents a torsion element of the Witt group $W^1(k)$ if and only if the signature of $\phi$ is zero with respect to every ordering of $k$.*

## 1. Artin-Schreier Theory

It will be convenient to speak not of the orderings of a field, as in theorem 7.2, but of real closures. We explain briefly the equivalence between the two ideas. The reader is referred to Bochnak, Coste and Roy [8, Chapter 1], Milnor and Husemoller [70, Chapter III §2] or Scharlau [93, p113] for further details.

Let us first recall a definition of ordering for a field:

DEFINITION 7.3. *An ordering of a field $k$ is a subset $\omega \subset k$ which is closed under addition and multiplication and is such that $k$ is the disjoint*

union:
$$k = -\omega \sqcup \{0\} \sqcup \omega$$
where $-\omega = \{-x \mid x \in \omega\}$.

Given an ordering $\omega$ one can define a relation $<_\omega$ on $k$ which satisfies all the usual axioms of an ordered set and moreover respects addition and multiplication. Let
$$x <_\omega y \quad \text{if and only if} \quad y - x \in \omega.$$
Inversely, given a relation $<$ one may set $\omega = \{x \in k \mid x > 0\}$.

If a field $k$ admits an ordering, the non-zero squares are certainly positive with respect to the ordering so $k$ is real. Conversely, in a real field the subset of elements which can be expressed as a sum of non-zero squares does not contain zero and can be extended, using Zorn's lemma, to an ordering $\omega$. Note that $\omega$ is not in general unique.

DEFINITION 7.4. A real field $k$ is said to be *real closed* if no proper algebraic extensions of $k$ are real. A *real closure* of a real field $k$ is an algebraic extension which is real closed.

A real closed field has much in common with the field $\mathbb{R}$ of real numbers. In particular, if $k$ is real closed then $k(\sqrt{-1})$ is algebraically closed and $k$ admits the unique ordering $\omega = \{x^2 \mid x \in k^\bullet\}$.

A Zorn's lemma argument shows that every ordered field $(k, \omega)$ has a real closure $k_\omega$ (whose unique ordering extends $\omega$). Conversely,

THEOREM 7.5 (Artin-Schreier). *There is precisely one real closure of an ordered field up to $k$-algebra isomorphism. Thus there is a canonical bijective correspondence between the orderings $\omega$ of $k$ and the $k$-isomorphism classes of real closures $k_\omega$ of $k$.*

We shall need one other basic fact about real fields:

LEMMA 7.6. *If a field $k$ is real but not real closed then either there exists an extension field of odd degree $q > 1$ or, for some ordering of $k$, there exists a positive non-square element $a \in k$ (or both).*

PROOF. See [**8**, p9] or [**93**, p113]. □

The main theorem of the chapter says that every element of infinite order in $W^\epsilon(R\text{--}k)$ is non-zero in some $W^\epsilon(R\text{--}k_\omega)$:

THEOREM 7.7. *Let $k$ be a (commutative) field with trivial involution.*
*a) If $k$ is not real then $2^n W^\epsilon(R\text{--}k) = 0$ for some $n$.*
*b) Suppose $k$ is real and $\{k_\omega\}_{\omega \in \Omega}$ is the family of real closures of $k$. Then every element $v$ in the kernel of the canonical map*
$$W^\epsilon(R\text{--}k) \to \bigoplus_\omega W^\epsilon(R\text{--}k_\omega)$$
*satisfies $2^n v = 0$ for some $n$.*
*c) Suppose $K = k(\sqrt{-1})$ is a field extension of degree 2 with involution fixing precisely $k$. Then every $v \in \mathrm{Ker}(W^\epsilon(R\text{--}k) \to W^\epsilon(R\text{--}K))$ satisfies $2v = 0$.*

In the particular case $k = \mathbb{Q}$ there is just one real closure, the set of real algebraic numbers $\mathbb{Q}_\omega = \mathbb{R} \cap \overline{\mathbb{Q}}$. Theorem 7.7 implies that the kernels of the natural maps
$$W^\epsilon(R\text{--}\mathbb{Q}) \to W^\epsilon(R\text{--}(\mathbb{R} \cap \overline{\mathbb{Q}}^-)) \to W^\epsilon(R\text{--}\overline{\mathbb{Q}}^-)$$
are 2-primary. Theorem 7.1 follows easily, because the natural map
$$W^\epsilon(R\text{--}\overline{\mathbb{Q}}^-) \cong \mathbb{Z}^{\oplus \overline{\mathcal{M}}^s(R\text{--}\overline{\mathbb{Q}}^-)} \to \mathbb{Z}^{\overline{\mathcal{M}}^s(R\text{--}\mathbb{C}^-)} \cong W^\epsilon(R\text{--}\mathbb{C}^-)$$
is an injection.

Note that when $R = \mathbb{Z}$ and $\epsilon = 1$, the kernel of $W^1(\mathbb{Q}) \to W^1(\mathbb{C}^-)$ already fails to be finitely generated; further discussion of invariants which detect the torsion part of $W^\epsilon(R\text{--}\mathbb{Q})$ can be found in chapter 11. The remainder of this chapter is devoted to the proof of theorem 7.7.

## 2. Frobenius reciprocity

NOTATION 7.8. Let $k$ denote any field with involution. For each involution-invariant element $a = \bar{a} \in k$ there is a one-dimensional (symmetric or) hermitian form
$$\phi : k \to k^*; x \mapsto (y \mapsto xa\bar{y})$$
which is denoted $\langle a \rangle_k$ or simply $\langle a \rangle$. Any (finite-dimensional) hermitian form over $k$ can be expressed as a direct sum of one-dimensional forms (diagonalized). One writes
$$\langle a_1, a_2, \cdots, a_m \rangle := \langle a_1 \rangle \oplus \langle a_2 \rangle \oplus \cdots \langle a_m \rangle.$$
We use the same notation to denote the Witt class in $W^1(k)$ represented by a non-singular hermitian form.

Let $K/k$ be an extension of commutative fields with involution. We have seen in chapter 3, section 3.2 that the inclusion of $k$ in $K$ induces a homomorphism of Witt rings $W^1(k) \to W^1(K)$ and a group homomorphism $W^\epsilon(R\text{--}k) \to W^\epsilon(R\text{--}K)$. Each of these maps is denoted $v \mapsto v_K$.

Suppose $s : K \to k$ is any $k$-linear map which respects the involutions. Then $s$ induces group homomorphisms
$$W^\epsilon(K) \to W^\epsilon(k) \quad \text{and} \quad W^\epsilon(R\text{--}K) \to W^\epsilon(R\text{--}k)$$
via the equation $s(M, \phi) = (M, s\phi)$ where
$$(s\phi)(m_1)(m_2) = s(\phi(m_1)(m_2))$$
for all $m_1, m_2 \in M$.

LEMMA 7.9 (Frobenius Reciprocity, Scharlau [92]). *Let $u \in W^1(K)$ and $v \in W^\epsilon(R\text{--}k)$. Then*
$$s(u.v_K) = su.v$$
*In particular,*
$$s(v_K) = (s\langle 1 \rangle_K).v$$

PROOF. Straightforward. □

## 3. Proof of Theorem 7.7

a) If $k$ is not formally real then $2^n W^1(k) = 0$ for some positive integer $n$ (e.g. see Milnor and Husemoller [**70**, p68 or p76]). Since $W^\epsilon(R\text{–}k)$ is a $W^\epsilon(k)$-module we have $2^n W^\epsilon(R\text{–}k) = 0$.

b) Assume the contrary, that there exists $v \in W^\epsilon(R\text{–}k)$ such that $2^n v \neq 0$ for all $n \geq 0$ and the image of $v$ is zero in each group $W^\epsilon(R\text{–}k_\omega)$. By Zorn's lemma there exists a maximal algebraic extension $K/k$ such that $2^n v_K \neq 0$ for all $n \geq 0$. Now $K$ is real by a) but is not real closed so lemma 7.6 says that either there exists an extension $K' = K(\xi)/K$ of odd degree $d > 1$ or for some ordering of $K$ there is a positive non-square $a \in K$.

In the former case, define $s : K' \to K$ by $s(1) = 1$ and $s(\xi^i) = 0$ for $1 \leq i \leq q-1$. Then $s\langle 1 \rangle_{K'} = \langle 1 \rangle_K \in W^1(K)$ because the elements $\xi^{\frac{q+1}{2}}, \cdots, \xi^{q-1}$ span a sublagrangian for $s(1_{K'})$ - see Scharlau [**93**, p49] for further details. Frobenius reciprocity gives the equation $s(v_{K'}) = v_K$. It follows from the maximality of $K$ that $2^n v_{K'} = 0$ for some $n$ so $2^n v_K = s(2^n v_{K'}) = 0$ and we have reached a contradiction.

The latter case is similar: Suppose $a \in K$ is non-square and positive in some ordering so $-a$ is also non-square. Let $K' = K(\sqrt{a})$ and define $s : K' \to K$ by $s(1) = 1$ and $s(\sqrt{a}) = 0$. Then $s\langle 1 \rangle_{K'} = \langle 1, a \rangle_K$ so Frobenius reciprocity gives $s(v_{K'}) = \langle 1, a \rangle . v_K$. Hence

$$(25) \qquad 2^n \langle 1, a \rangle . v_K = 0 \text{ for some } n.$$

Now $-a$ is another non-square in $K$ so by the same argument

$$(26) \qquad 2^n \langle 1, -a \rangle . v_K = 0 \text{ for some } n.$$

For suitably large $n$ we may sum equations (25) and (26) to find that $2^{n+1} v_K = 0 \in W^\epsilon(R\text{–}K)$. Once again we have reached a contradiction.

c) Let $v \in W^\epsilon(R\text{–}k)$ and suppose $v_K = 0$. Define $s : K \to k$ by $s(1) = 1$ and $s(\sqrt{-1}) = 0$. Frobenius reciprocity yields

$$0 = s(v_K) = \langle 1, 1 \rangle . v = 2v \ .$$

CHAPTER 8

# Characters

Having associated cobordism invariants $\sigma_{M,b}$ to simple self-dual representations of the ring $P_\mu$ we turn now to the theory of characters to distinguish such representations. A 'trace' invariant for boundary links was first introduced by Farber [**25**] under the assumption that the ground ring should be an algebraically closed field. Some of his work was later simplified by Retakh, Reutenauer and Vaintrob [**85**].

In the present chapter $k$ is any field of characteristic zero. All representations will be implicitly assumed to be finite-dimensional. The characters of simple representations of an arbitrary associative ring $R$ over $k$ are shown to be linearly independent which implies that a semisimple representation is determined up to isomorphism by its character.

Recalling from chapter 5 that $F_\mu$-link signatures are associated to self-dual representations, it is shown at the conclusion of the present chapter that a semisimple representation is self-dual if and only if the corresponding character $\chi : R \to k$ respects the involutions of $R$ and $k$ (cf [**25**, §6]).

The books by Serre [**94**] and Curtis and Reiner [**19**] on representation theory of finite groups are used for basic reference. We do not assume the invariant theory of Procesi and Le Bruyn which we employed in chapter 6.

The linear independence of characters will be exploited again in the study of rationality questions in chapter 9 below.

## 1. Artin algebras

We must first recall the definition and basic theory of Artin algebras

DEFINITION 8.1. A $k$-algebra $S$ is called *Artinian* or an *Artin $k$-algebra* if every descending chain of left ideals

$$I_1 \supset I_2 \supset \cdots \supset I_n \supset \cdots$$

terminates, i.e. eventually, $I_n = I_{n+1} = I_{n+2} = \cdots$.

For example, every finite-dimensional $k$-algebra is Artinian. Recall that a ring $S$ is by definition *simple* if $0$ and $S$ are the only two-sided ideals.

THEOREM 8.2. *i) An Artin $k$-algebra $S$ is simple if and only if it admits a faithful simple representation $\rho : S \hookrightarrow \mathrm{End}_k(M)$.*
*ii) (Skolem-Noether) A simple Artin $k$-algebra admits a unique simple representation.*

## 2. Independence of Characters

DEFINITION 8.3. Suppose $A$ is a commutative ring. The *character* $\chi_M \in \mathrm{Hom}_{\mathbb{Z}}(R, A)$ of a representation $(M, \rho)$ of $R$ over $A$ is
$$\chi_M : R \to A$$
$$r \mapsto \mathrm{Trace}(\rho(r)).$$

We concentrate here on the case $A = k$ where $k$ denotes a field of characteristic zero.

DEFINITION 8.4. A representation $M$ over $k$ is called *simple* or *irreducible* if there are no subrepresentations other than 0 and $M$. One says $M$ is *semisimple* if it is a direct sum of simple representations.

PROPOSITION 8.5. *If* $(M_1, \rho_1), (M_2, \rho_2), \cdots, (M_l, \rho_l)$ *are (pairwise) non-isomorphic simple representations then* $\chi_{M_1}, \chi_{M_2}, \cdots, \chi_{M_l}$ *are linearly independent over* $k$.

COROLLARY 8.6. *Let* $k$ *be a (commutative) field of characteristic zero. Two semisimple representations* $(M, \rho)$ *and* $(M', \rho')$ *of* $R$ *over* $k$ *are isomorphic if and only if* $\chi_M = \chi_{M'}$.

We shall prove proposition 8.5 by studying the kernels of the maps $\rho_i$. Suppose $A$ is a ring, $M$ is an $A$-module and $T$ is an abelian group. A homomorphism $f \in \mathrm{Hom}_{\mathbb{Z}}(T, M)$ determines, and is determined by, the $A$-module map $Af : A \otimes_{\mathbb{Z}} T \to M$ given by $Af(a \otimes t) = af(t)$ for all $a \in A$ and $t \in T$. There is a commutative diagram

$$\begin{array}{ccc} T & \xrightarrow{f} & M \\ {\scriptstyle i} \downarrow & \nearrow {\scriptstyle Af} & \\ A \otimes_{\mathbb{Z}} T & & \end{array}$$

where $i$ is the natural map $t \mapsto 1 \otimes t$. In particular, a representation $\rho : R \to \mathrm{End}_k(M)$ and its character $\chi_M : R \to k$ determine and are determined by $k$-linear maps $k\rho : k \otimes_{\mathbb{Z}} R \to \mathrm{End}_k(M)$ and $k\chi_M : k \otimes_{\mathbb{Z}} R \to k$ respectively.

LEMMA 8.7. *Let* $R$ *be any ring. Two simple representations* $(M, \rho)$ *and* $(M', \rho')$ *over* $k$ *are isomorphic if and only if* $\mathrm{Ker}(k\rho) = \mathrm{Ker}(k\rho')$.

PROOF. If $\theta : (M, \rho) \to (M', \rho')$ is an isomorphism then
$$k\rho(x) = \theta^{-1} k\rho'(x) \theta$$
for all $x \in k \otimes_{\mathbb{Z}} R$ so $\mathrm{Ker}(k\rho) = \mathrm{Ker}(k\rho')$.

Conversely, suppose $\mathrm{Ker}(k\rho) = \mathrm{Ker}(k\rho') = I$. The quotient $\frac{k \otimes_{\mathbb{Z}} R}{I}$ has faithful irreducible representations $(M, k\rho)$ and $(M', k\rho')$. Since $\frac{k \otimes_{\mathbb{Z}} R}{I}$ is finite-dimensional over $k$, and hence Artinian it follows from theorem 8.2 that $(M, k\rho) \cong (M', k\rho')$, so $(M, \rho) \cong (M', \rho')$. $\square$

PROOF OF PROPOSITION 8.5. Let $I_i = \mathrm{Ker}(\rho_i) \subset k \otimes_{\mathbb{Z}} R$ for $1 \le i \le l$. By lemma 8.7 and theorem 8.2 i), $I_1, \cdots, I_l$ are distinct maximal two-sided ideals of $k \otimes_{\mathbb{Z}} R$. In particular, $I_i + I_j = R$ for $i \ne j$ so, fixing $i \in \{1, \cdots, l\}$,

$$R = \prod_{j \mid j \ne i} (I_i + I_j) \subset I_i + \prod_{j \mid j \ne i} I_j$$

and there is an equation $1 = x'_i + x_i$ with $x'_i \in I_i$ and $x_i \in \prod_{j \mid j \ne i} I_j \subset \bigcap_{j \mid j \ne i} I_j$. Thus $x_i$ acts as the identity on $M_i$ and as zero on the other $M_j$ (cf [**19**, ex9 p170]) so $(k\chi_{M_i})(x_i) = \dim_k M_i$ and $(k\chi_{M_j})(x_i) = 0$ when $j \ne i$.

If there is a linear relation $\sum_{j=1}^{l} a_j \chi_{M_j} = 0$ with each $a_j \in k$ then, evaluating at $x_i$, we have

$$0 = \sum_{j=1}^{l} a_j (k\chi_{M_j})(x_i) = a_i \dim_k M_i$$

so $a_i = 0$ for each $i \in \{1, \cdots, l\}$. □

**2.1. Self-Dual Characters.** Let $R$ be a ring with involution and let $k$ be a field with involution. Recall from example 3.16 the duality functor $(M, \rho) \mapsto (M^*, \rho^*)$ on the category $(R\text{–}k)$-Proj.

LEMMA 8.8. *A semisimple representation $M$ of $R$ over $k$ is self dual $(M, \rho) \cong (M^*, \rho^*)$ if and only if the character $\chi_M$ respects the involutions, i.e. $\chi_M(\bar{r}) = \overline{\chi_M(r)}$ for all $r \in R$.*

PROOF. By corollary 8.6, $(M, \rho) \cong (M^*, \rho^*)$ if and only if $\chi_M = \chi_{M^*}$. Now $\chi_{M^*}(\bar{r}) = \mathrm{Trace}(\rho^*(\bar{r})) = \mathrm{Trace}(\rho(r)^*) = \overline{\mathrm{Trace}(\rho(r))} = \overline{\chi_M(r)}$ so $(M, \rho) \cong (M^*, \rho^*)$ if and only if $\chi_M(\bar{r}) = \overline{\chi_M(r)}$ for all $r \in R$. □

# CHAPTER 9

# Detecting Rationality and Integrality

As discussed in chapter 2, and in more detail in chapter 5, the composite

$$W^\epsilon(P_\mu\text{-}\mathbb{Z}) \to W^\epsilon(P_\mu\text{-}\mathbb{C}^-) \cong \mathbb{Z}^{\oplus \infty}$$

associates to a Seifert form one signature invariant $\sigma_{M,b}$ for each self-dual simple complex representation $M$ of the quiver $P_\mu$.

These are a complete set of signatures, but some are surplus to requirements. One source of redundancy derives from the fact that each signature is equal to its complex conjugate $\sigma_{M,b} = \sigma_{\overline{M},\overline{b}}$. A second source is that a necessary condition for a signature $\sigma_{M,b}$ to be non-zero is that $M$ should be *algebraically integral*, i.e. $M$ should be a summand of some representation induced up from an integral representation.

The present chapter shows how the character can be used to identify the algebraically integral representations among a variety of complex representations. The chapter is arranged in two sections dubbed rationality and integrality. In the rationality section we study an arbitrary field extension $k_0 \subset k$ in characteristic zero showing that a semisimple representation $M$ of an arbitrary associative ring $R$ over $k$ is a summand of a representation induced up from $k_0$ if and only if the character $\chi_M$ takes values in a finite extension field of $k_0$.

Turning to questions of integrality, we prove that a representation $M$ over $\mathbb{C}$ is a summand of an integral representation if and only if $\chi_M$ takes values in the ring of algebraic integers of some finite extension of $\mathbb{Q}$. In particular, when $R$ is a path ring of a quiver, one need only check that the traces of oriented cycles lie in such an algebraic number ring.

## 1. Rationality

In this section $k_0 \subset k$ denotes a (possibly infinite) extension of characteristic zero fields.

**1.1. Induction.** If $M_0$ is a representation of $R$ over $k_0$ then, as in the third paragraph of chapter 3, section 1, $k \otimes_{k_0} M$ is a representation of $R$ over $k$. A representation over $k$ which is isomorphic to $k \otimes_{k_0} M_0$ for some $M_0$ is called a $k_0$-*induced representation*, or simply an *induced representation* when the identity of $k_0$ is clear.

LEMMA 9.1 (Induction Lemma). *Suppose* $(M_0, \rho_0)$ *and* $(M_0', \rho_0')$ *are representations of $R$ over $k_0$.*

i) $M_0$ is semisimple if and only if $k \otimes_{k_0} M_0$ is semisimple.

ii) If $M_0$ and $M_0'$ are semisimple then
$$M_0 \cong M_0' \iff k \otimes_{k_0} M_0 \cong k \otimes_{k_0} M_0'.$$

iii) If $M_0$ and $M_0'$ are non-isomorphic simple representations, no summand of $k \otimes_{k_0} M_0$ is isomorphic to a summand of $k \otimes_{k_0} M_0'$.

PROOF. i) See for example Curtis and Reiner [**19**, 3.56(iii), 7.5] or Bourbaki [**9**, Theorem 2 p87].

ii) Follows from proposition 8.6 since $\chi_{k \otimes_{k_0} M_0} = \chi_{M_0} \in \text{Hom}_{\mathbb{Z}}(R, k_0)$.

iii) As in the proof of proposition 8.6 there exists an element $x \in k_0 \otimes_{\mathbb{Z}} R$ such that $x$ acts as the identity on $M_0$ and as zero on $M_0'$. Now $x$ also acts as the identity on $k \otimes_{k_0} M_0$ and as zero on $k \otimes_{k_0} M_0'$. □

In the following lemma, we assume that $R$, $k_0$ and $k$ are endowed with involutions.

LEMMA 9.2. *Suppose $M_0$ is simple and $M$ is a simple summand of the induced representation $k \otimes_{k_0} M_0$. If $M$ is self-dual then $M_0$ is also self-dual.*

PROOF. Observing that $M^*$ is a summand of $k \otimes_{k_0} M_0^*$, if $M \cong M^*$ then, lemma 9.1 iii) implies that $M_0 \cong M_0^*$. □

The converse to lemma 9.2 is not true in general; a self-dual simple representation $M_0$ over $k_0$ decomposes over $k$ into summands among which there may be dual pairs $M \oplus M^*$.

**1.2. Restriction.** If $k$ is a finite-dimensional extension field of $k_0$ and $(M, \rho)$ is a representation of $R$ over $k$ then $M$ can be regarded as a representation over $k_0$:

DEFINITION 9.3. The *restriction* $\text{Res}_{k_0}^k (M, \rho)$ of $M$ to $k_0$ is the representation $(\text{Res}_{k_0}^k M, F \circ \rho)$ where $\text{Res}_{k_0}^k M$ is $M$ regarded as a vector space over $k_0$ and $F : \text{End}_k M \to \text{End}_{k_0} M$ is the forgetful map.

LEMMA 9.4 (Restriction Lemma). *Suppose $k$ is a finite-dimensional extension of $k_0$ and $M$ is a representation of $R$ over $k$. Then $M$ is semisimple if and only if the restriction $\text{Res}_{k_0}^k M$ is semisimple.*

PROOF. Suppose first that $M$ is simple and let $N_0$ be any simple submodule of $\text{Res}_{k_0}^k M$. If $a_1, \cdots, a_m$ is a basis for $k$ over $k_0$ then for each $i$, $a_i N_0$ is a simple representation of $R$ over $k_0$ and $\sum_{i=1}^m a_i N_0 = \text{Res}_{k_0}^k M$. Hence $\text{Res}_{k_0}^k M$ is semisimple. It follows immediately that if $M$ is semisimple then $\text{Res}_{k_0}^k M$ is semisimple.

Conversely, suppose $\text{Res}_{k_0}^k M$ is semisimple. There is a natural surjection $k \otimes_{k_0} \text{Res}_{k_0}^k M \twoheadrightarrow M$ and by lemma 9.1 i) above $k \otimes_{k_0} \text{Res}_{k_0}^k M$ is semisimple. □

## 1. RATIONALITY

### 1.3. Criteria for Rationality.

PROPOSITION 9.5. *Suppose $k_0 \subset k$ and $(M, \rho)$ is a semisimple representation of $R$ over $k$. The following are equivalent:*

(1) *There exists a positive integer $d$ and a semisimple representation $M_0 = (M_0, \rho_0)$ such that $M^{\oplus d} \cong k \otimes_{k_0} M_0$.*
(2) $\det(x - \rho(r)) \in k_0[x]$ *for all $r \in R$.*
(3) *The character $\chi_M$ is $k_0$-valued, i.e. $\mathrm{Trace}(\rho(r)) \in k_0$ for all $r \in R$.*

REMARK 9.6. Case a) of the proof below, together with lemma 9.1 iii), demonstrates that if $M$ is a simple representation over $k$ and $\chi_M$ is $k_0$-valued then $M_0$ can also be chosen, in a unique way, to be a simple representation. The positive integer $d = d(M, k/k_0)$ is then the 'relative Schur index'.

PROOF OF PROPOSITION 9.5. $2 \Leftrightarrow 3$: Let $f = \det(1 - x\rho(r)) \in k[x]$ be the 'reverse characteristic polynomial' of $\rho(r)$ which has the property $f \in k_0[x]$ if and only if $\det(x - \rho(r)) \in k_0[x]$. The exponential trace formula

$$\left(-\frac{d}{dx}\log f =\right) - f^{-1}\frac{df}{dx} = \sum_{n \geq 1} \mathrm{Trace}(\rho(r^n))x^{n-1}$$

in the ring $k[[x]]$ of formal power series implies that $f$ has coefficients in $k_0$ if and only if $\mathrm{Trace}(\rho(r^i)) \in k_0$ for all $i \geq 1$.

$1 \Rightarrow 3$: $\chi_M = \frac{1}{d}\chi_{M^{\oplus d}} = \frac{1}{d}\chi_{k \otimes_{k_0} M_0} = \frac{1}{d}\chi_{M_0}$ so $\chi_M$ is $k_0$-valued.

The rest of the proof deduces statement 1 from statements 2 and 3. We proceed in stages assuming that a) $M$ is simple and b) $M$ is semisimple. In case b) we first consider finite extensions $k/k_0$ before addressing arbitrary field extensions (in characteristic zero).

a) Suppose that $\chi_M$ is $k_0$-valued and $M$ is simple. Let $R$ act on $\mathrm{End}_k M$ by left multiplication $r.\alpha = \rho(r)\alpha$ so that, as a representation, $\mathrm{End}_k M$ is isomorphic to $M^{\oplus m}$ where $m = \dim_k M$. Similarly, $k\rho(R)$ and $k_0\rho(R)$ can be regarded as representations of $R$ over $k$ and $k_0$ respectively.

If we can prove that the natural surjection

$$(27) \qquad k \otimes_{k_0} k_0\rho(R) \to k\rho(R) \subset \mathrm{End}_k M \cong M^{\oplus n}$$

is an isomorphism, i.e. that $\dim_{k_0} k_0\rho(R) = \dim_k k\rho(R)$, then any simple subrepresentation $M_0$ of $k_0\rho(R)$ has the property $k \otimes_{k_0} M_0 \cong M^{\oplus d}$ for some $d$.

To show that (27) is an isomorphism, let $S \subset R$ be a finite subset with the property that $\rho(S)$ is a basis for $k\rho(R)$ over $k$. Note that $M$ is a faithful simple representation of $k\rho(R)$ so $k\rho(R)$ is a simple $k$-algebra by theorem 8.2 i). It follows that

$$\Phi : k\rho(R) \to k^S$$
$$\alpha \mapsto (s \mapsto \mathrm{Trace}(\rho(s)\alpha))$$

is injective, for if $\mathrm{Trace}(\rho(s)\alpha) = 0$ for all $s \in S$ then $\mathrm{Trace}(k\rho(R)\alpha) = 0$ which implies that $\mathrm{Trace}(k\rho(R)\alpha k\rho(R)) = 0$ and hence that $\alpha = 0$. Since

$\dim(k\rho(R)) = \dim(k^S)$, $\Phi$ is an isomorphism of $k$-vector spaces and the restriction $\Phi| : k_0\rho(R) \to k_0^S$ is an isomorphism of $k_0$-vector spaces. It follows that $\dim_{k_0} k_0\rho(R) = |S| = \dim_k k\rho(R)$ so (27) is an isomorphism as required.

b) Suppose $k$ is finite-dimensional over $k_0$. Let $M_0 = \mathrm{Res}^k_{k_0} M$ (cf Serre [**94**, lemma 12 p92]). Now

$$\chi_{k\otimes_{k_0} M_0} = \chi_{M_0} = \mathrm{Tr}_{k/k_0}(\chi_M) = [k : k_0]\chi_M$$

since $\chi_M$ is $k_0$-valued. By proposition 8.6 $k \otimes_{k_0} M_0 \cong M^{\oplus [k:k_0]}$.

If $k/k_0$ is an arbitrary extension, the following lemma reduces the problem to the finite case:

LEMMA 9.7. *Let $k/k_0$ be an arbitrary field extension (in characteristic zero). Suppose $\chi_1, \chi_2, \cdots, \chi_l$ are characters of simple representations over $k$, and $\sum_{i=1}^l t_i\chi_i$ is $k_0$-valued with each $t_i \in \mathbb{Z}$. Then there exists a finite extension $k_1/k_0$ such that $\chi_i$ is $k_1$-valued for $1 \leq i \leq l$.*

Suppose $M$ is a semisimple representation over $k$ and $\chi_M$ is $k_0$-valued. Let us deduce from lemma 9.7 that $M^{\oplus d} \cong k \otimes_{k_0} M_0$. Writing $M$ as a direct sum of simple representations

$$M = (M^{(1)})^{\oplus t_1} \oplus \cdots \oplus (M^{(l)})^{\oplus t_l}$$

we obtain an equation $\chi_M = \sum_{i=1}^l t_i\chi_{M^{(i)}}$. By lemma 9.7, all the simple characters $\chi_{M^{(i)}}$ take values in some finite extension $k_1$ of $k_0$ and by case a) above, there exists a positive integer $d^{(i)}$ and a representation $M_1^{(i)}$ over $k_1$ such that $(M^{(i)})^{\oplus d^{(i)}} \cong k \otimes_{k_1} M_1^{(i)}$. Writing $d_1 = \mathrm{lcm}(d^{(1)}, \cdots, d^{(l)})$ we have $M^{\oplus d_1} \cong k \otimes_{k_1} M_1$ for some semisimple representation $M_1$ over $k_1$. As in b) above $M_1^{\oplus [k_1:k_0]} \cong k_1 \otimes_{k_0} M_0$ where $M_0 = \mathrm{Res}^{k_1}_{k_0} M_1$ so $M^{\oplus d_1[k_1:k_0]} \cong k \otimes_{k_0} M_0$.

PROOF OF LEMMA 9.7. The proof is divided into three stages, which apply to increasingly general classes of extensions $k/k_0$. We consider i) Galois extensions; ii) Algebraic extensions; iii) Arbitrary extensions.

i) The Galois group $G = \mathrm{Gal}(k/k_0)$ acts on $\mathrm{Hom}_\mathbb{Z}(R, k)$ by the equation $\chi^g(r) = g\chi(r)$ where $g \in G$, $\chi \in \mathrm{Hom}_\mathbb{Z}(R, k)$ and $r \in R$.

If $\chi = t_1\chi_1 + \cdots + t_l\chi_l$ is $k_0$-valued then

$$t_1\chi_1 + \cdots + t_l\chi_l = \chi = \chi^g = t_1\chi_1^g + \cdots t_l\chi_l^g$$

so, by the linear independence of characters (proposition 8.5), $G$ permutes the set $\{\chi_1, \cdots, \chi_l\}$.

The kernel of this action is a normal subgroup $H \leq G$ of finite index. Let $k_1 = k^H$ be the intermediate field $k_0 \subset k_1 \subset k$ of elements fixed by $H$. Each character $\chi_i$ is fixed by $H$ and is therefore $k_1$-valued so it remains to note that $[k_1 : k_0] \leq [G : H] < \infty$ by the following standard Galois theory argument: If $\alpha \in k_1$ then the orbit $G.\alpha \subset k_1$ is finite with cardinality at most $[G : H]$, so $\alpha$ is a root of a polynomial $\prod_{\lambda \in G\alpha}(x - \lambda) \in k_0[x]$ of degree at most $[G:H]$. It follows by the primitive element theorem I.3 that $k_1$ is a finite extension of $k_0$ with $[k_1:k_0] \leq [G:H]$. (In fact $[k_1:k_0]=[G:\overline{H}]$ where $\overline{H}$ is the closure of $H$ in the Krull topology - see for example [86, Theorem 2.11.3]).

ii) Given an algebraic extension $k/k_0$, let $K/k_0$ be a Galois extension such that $K \supset k$ (e.g. let $K$ be the algebraic closure $\overline{k_0}$). By case i) there exists a finite field extension $K_1$ of $k_0$ such that $k_0 \subset K_1 \subset K$ and each $\chi_i$ is $K_1$-valued. Since $\chi_i$ is certainly $k$-valued we may set $k_1 = k \cap K_1$.

iii) Arguing as in ii) we can assume that $k$ is algebraically closed and therefore contains the algebraic closure $\overline{k_0}$ of $k_0$. Let $(M^{(i)}, \rho^{(i)})$ be a simple representation over $k$ with character $\chi_i$ and let $M = (M^{(1)})^{\oplus t_1} \oplus \cdots \oplus (M^{(l)})^{\oplus t_l}$. Suppose $\chi_M$ is $k_0$-valued. For each $r \in R$ the characteristic polynomial of the endomorphism $\rho^{(1)}(r)^{\oplus t_1} \oplus \cdots \oplus \rho^{(l)}(r)^{\oplus t_l}$ is in $k_0[x]$ so the eigenvalues are in $\overline{k_0}$. It follows that the eigenvalues of each $\rho^{(i)}(r)$ are in $\overline{k_0}$ so $\chi_i$ is $\overline{k_0}$-valued for all $i$ and we have reduced the problem to case i). □

This completes the proof of proposition 9.5. □

REMARK 9.8. For the purposes of our application to boundary links one could add the hypothesis that $R$ is finitely generated, which simplifies the proof of lemma 9.7 as follows: If $k/k_0$ is an algebraic extension and $\chi$ is the character of a representation $M$ over $k$ then, fixing a basis for $M$, a finite generating set for $R$ acts via (finitely many) square matrices, the entries in which generate a finite extension field $k_1$ of $k_0$. Plainly $\chi$ is $k_1$-valued.

COROLLARY 9.9. *Under the hypotheses of proposition 9.5, the following are equivalent:*

(1) *There exists a semisimple representation $(M_0, \rho_0)$ over $k_0$ such that $M$ is a summand of $k \otimes_{k_0} M_0$.*
(2) *There exists a finite extension $k_1/k_0$, a positive integer $d$ and a semisimple representation $(M_1, \rho_1)$ over $k_1$ such that $M^{\oplus d} \cong k \otimes_{k_1} M_1$.*
(3) *There exists a finite extension $k_1/k_0$ such that $\det(x - \rho(r)) \in k_1[x]$ for all $r \in R$.*
(4) *There exists a finite extension $k_1/k_0$ such that the character $\chi_M$ is $k_1$-valued.*

PROOF. Statements 2, 3 and 4 are equivalent by proposition 9.5.
$1 \Rightarrow 3$: Suppose $M \oplus M' \cong k \otimes_{k_0} M_0$. Let $\chi_1, \cdots, \chi_l$ denote the characters of the distinct isomorphism classes of simple summands of $M \oplus M'$. By lemma 9.7 above there is a finite extension $k_1$ of $k_0$ such that all the characters $\chi_i$ are $k_1$-valued. Hence $\chi_M$ is $k_1$-valued.
$2 \Rightarrow 1$: It suffices to show that there exists a semisimple representation $M_0$ over $k_0$ such that $M_1$ is a summand of $k_1 \otimes_{k_0} M_0$. Set $M_0 = \operatorname{Res}_{k_0}^{k_1} M_1$. By lemma 9.4 $M_0$ is semisimple and there is a natural surjection $k_1 \otimes_{k_0} M_0 \twoheadrightarrow M_1$. □

DEFINITION 9.10. A semisimple complex representation $M$ of $R$ is *algebraic* if $M$ is a summand of a $\mathbb{Q}$-induced representation.

COROLLARY 9.11. *A semisimple complex representation $M$ is algebraic if and only if $\chi_M$ takes values in an algebraic number field.*

PROOF. This is a special case of corollary 9.9. □

DEFINITION 9.12. Two algebraic representations $M$ and $M'$ are *conjugate* if both $M$ and $M'$ are summands of the same $\mathbb{Q}$-induced representation.

COROLLARY 9.13. *i) Two simple algebraic representations $M$ and $M'$ are conjugate if and only if $\chi_M$ and $\chi_{M'}$ lie in the same $\mathrm{Gal}(\overline{\mathbb{Q}}/\mathbb{Q})$-orbit.*
*ii) If $R$ is the path ring of a quiver $Q$, then conjugate simple algebraic representations have the same dimension vector.*

PROOF. i) Let $G = \mathrm{Gal}(\overline{\mathbb{Q}}/\mathbb{Q})$. If $O$ denotes the $G$-orbit of $\chi_M$ then $\sum_{\chi \in O} \chi$ is $G$-invariant and therefore $\mathbb{Q}$-valued. Note that $O$ is finite by corollary 9.9. By proposition 9.5, there is a positive integer $d$ such that $d \sum_{\chi \in O} \chi$ is the character of a rational representation $M_0$. Every rational subrepresentation has a $G$-invariant character which, by proposition 8.5, must be $d' \sum_{\chi \in O} \chi$ for some integer $d' \leq d$. If we choose $d$ to be minimal then $M_0$ is simple Lemma 9.1 iii) implies that $M'$ is conjugate to $M$ if and only if $\chi_{M'} \in O$.

ii) Recall that $e_x$ denotes the idempotent in $\mathbb{Z}Q$ corresponding to the trivial path at $x \in Q_0$. After i) it suffices to observe that for any semisimple complex representation $M$ of $Q$ and any vertex $x \in Q_0$

$$\dim_{\mathbb{C}}(M_x) = \chi(e_x) \in \mathbb{Z}_{\geq 0}$$

and is invariant under $\mathrm{Gal}(\overline{\mathbb{Q}}/\mathbb{Q})$. □

## 2. Integrality

In place of a field extension, we consider in this section the inclusion of $\mathbb{Z}$ in $\mathbb{Q}$. The general theory of integral representations is well-known to be far more subtle. In particular, $(R\text{–}\mathbb{Z})\text{-Proj}$ is not an abelian category. Nevertheless, one can give criteria for a rational representation to be induced from some integral representation:

PROPOSITION 9.14. *Suppose $(M, \rho)$ is a semisimple representation of $R$ over $\mathbb{Q}$. The following are equivalent:*
  (1) *There exists a representation $(M_0, \rho_0)$ of $R$ over $\mathbb{Z}$ with the property that $M \cong \mathbb{Q} \otimes_{\mathbb{Z}} M_0$.*
  (2) $\det(x - \rho(r)) \in \mathbb{Z}[x]$ *for all $r \in R$.*
  (3) *The character $\chi_M$ is $\mathbb{Z}$-valued.*

CAVEAT 9.15. In contrast to remark 9.6, even if one assumes that $M$ is simple, $M_0$ is not in general unique.

## 2. INTEGRALITY

PROOF OF PROPOSITION 9.14. $1 \Rightarrow 2$: Immediate.

$2 \Rightarrow 3$: Trace($\rho(r)$) is a coefficient of $\det(x - \rho(r))$.

$3 \Rightarrow 2$: We assume $\chi_M$ is $\mathbb{Z}$-valued and aim to prove that $f = \det(1 - x\rho(r))$ is $\mathbb{Z}$-valued. Once again we use the exponential trace formula

$$-f^{-1}\frac{df}{dx} = \sum_{i \geq 1} \text{Trace}(\rho(r^i)x^{i-1}) \in \mathbb{Z}[[x]].$$

If $f = 1 + a_1 x + a_2 x^2 + \cdots + a_n x^n$, let $b$ denote the smallest positive integer such that $ba_i \in \mathbb{Z}$ for all $i$, i.e. $b$ is the least common multiple of the denominators of the $a_i$. We aim to prove that $b = 1$. Now $f = 1 + g/b$ where $g = ba_1 x + \cdots + ba_n x^n \in A[x]$ and $\text{hcf}(b, g) = 1$ so

$$f^{-1} = 1 - \frac{g}{b} + \frac{g^2}{b^2} - \frac{g^3}{b^3} + \cdots$$

If $b$ has a prime factor $p$ then plainly $f^{-1}\frac{df}{dx} \notin \mathbb{Z}[[x]]$. Thus $b = 1$ and $f \in \mathbb{Z}[x]$ as required.

$2, 3 \Rightarrow 1$: Suppose first that $M$ is a simple representation so that $\mathbb{Q}\rho(R)$ is a simple algebra. Choose a minimal finite subset $S \subset R$ such that $\rho(S)$ is a $\mathbb{Q}$-basis for $\mathbb{Q}\rho(R)$. The vector space isomorphism

$$\mathbb{Q}\rho(R) \to \mathbb{Q}^S$$
$$\alpha \mapsto (s \mapsto \text{Trace}(\alpha\rho(s)))$$

restricts to an injection $\rho(R) \to \mathbb{Z}^S$, so $\rho(R)$ is finitely generated as a $\mathbb{Z}$-module. If $v_1, \cdots, v_n$ is a $\mathbb{Q}$-basis for $M$ then we may set $M_0 = \sum_{i=1}^n \rho(R)v_i$ and the natural map $\mathbb{Q} \otimes_{\mathbb{Z}} M_0 \to \mathbb{Q}M_0 = M$ is an isomorphism.

To extend the result to the semisimple case note that a direct sum $(M, \rho) \oplus (M', \rho')$ satisfies condition 2 if and only if $(M, \rho)$ and $(M', \rho')$ both satisfy condition 2; for

$$\det\left(x - (\rho(r) \oplus \rho'(r))\right) = \det\left(x - \rho(r)\right) \det\left(x - \rho'(r)\right)$$

is in $\mathbb{Z}[x]$ if and only if $\det(x - \rho(r)) \in \mathbb{Z}[x]$ and $\det(x - \rho'(r)) \in \mathbb{Z}[x]$. $\square$

DEFINITION 9.16. A complex representation $M$ of $R$ is *algebraically integral* if $M$ is a direct summand of a $\mathbb{Z}$-induced representation.

Let $\mathcal{O}$ denote the ring of algebraic integers. If $K$ is an algebraic number field, then $\mathcal{O}_K = \mathcal{O} \cap K$ denotes the ring of algebraic integers in $K$.

COROLLARY 9.17. *Suppose $M = (M, \rho)$ is a semisimple representation of $R$ over $\mathbb{C}$. Then the following are equivalent:*
  (1) *$M$ is algebraically integral.*
  (2) *There exists an algebraic number field $K$ such that $\det(x - \rho(r)) \in \mathcal{O}_K[x]$ for all $r \in R$.*
  (3) *There exists an algebraic number field $K$ such that $\chi_M$ is $\mathcal{O}_K$-valued.*

PROOF. 1 ⇒ 2: The eigenvalues of each endomorphism $\rho(r)$ are algebraic integers so each characteristic polynomial $\det(x - \rho(r))$ has coefficients in $\mathcal{O}$. Since $M$ is a summand of $\mathbb{C} \otimes_\mathbb{Q} \mathbb{Q} \otimes_\mathbb{Z} M_0$, corollary 9.9 implies that there is a finite extension $K$ of $\mathbb{Q}$ such that each polynomial $\det(x - \rho(r))$ has coefficients in $\mathcal{O}_K$.

2 ⇒ 3: Trivial.

3 ⇒ 1: Let $\Sigma$ be the set of embeddings $K \hookrightarrow \mathbb{C}$. If $\chi_M$ is $\mathcal{O}_K$-valued then $M$ is a summand of a complex representation $M'$ with the $\mathbb{Z}$-valued character $\bigoplus_{\sigma \in \Sigma} \chi_M^\sigma$. By proposition 9.5 $M$ is a summand of a $\mathbb{Q}$-induced representation $(M')^{\oplus d}$. By proposition 9.14, $(M')^{\oplus d}$ is $\mathbb{Z}$-induced. □

COROLLARY 9.18. *Let $M = (M, \rho)$ be a semisimple representation of a quiver $Q$. The following are equivalent:*

(1) *$M$ is algebraically integral.*
(2) *There is an algebraic number field $K$ such that the characteristic polynomial of every oriented cycle in the quiver has coefficients in $\mathcal{O}_K$, i.e. $\det(x - \rho(r)) \in \mathcal{O}_K[x]$.*
(3) *There is an algebraic number field $K$ such that $\operatorname{Trace}(\rho(r))$ lies in $\mathcal{O}_K$ for every oriented cycle $r \in Q$.*

PROOF. It follows immediately from corollary 9.17 that 1 ⇒ 2 and 2 ⇒ 3.

To show that 3 ⇒ 1, observe that if $r \in Q$ is a non-trivial path but not a cycle then $\operatorname{Trace}(\rho(r)) = 0$. Thus condition 3 of corollary 9.18 implies condition 3 of corollary 9.17. □

# CHAPTER 10

# Representation Varieties: Two Examples

In this chapter we describe the varieties of isomorphism classes of semisimple representations of $P_2$ for dimension vectors $\alpha = (1,1)$ and $\alpha = (1,2)$. In each case we compute the open subvariety of simple representations, the real variety of self-dual representations and the subset of algebraically integral representations. I am grateful to Raf Bocklandt for a helpful discussion of the case $\alpha = (1,2)$.

By theorem 6.4 the coordinate ring $\mathbb{C}[X]^{GL(\alpha)}$ for the variety of semisimple isomorphism classes with dimension vector $\alpha$ is generated by a finite number of traces of oriented cycles in the quiver $P_\mu$. Moreover the relations between these generators can all be deduced from 'Cayley-Hamilton relations' (for further explanation see Procesi [**79**]). While explicit sets of generators and relations can in principle be computed for any dimension vector $\alpha$ a general formula is not available at the time of writing.

A particular representation $(M, \rho)$ of $P_2$ can be displayed as follows:

(28) $$\left(\begin{array}{c|c} a_{11} & a_{12} \\ \hline a_{21} & a_{22} \end{array}\right) = a_{11} \bigcirc \bullet \overset{a_{21}}{\underset{a_{12}}{\rightleftarrows}} \bullet \bigcirc a_{22}$$

If $e_{ij}$ denotes the arrow from vertex $j$ to vertex $i$ in the quiver $P_\mu$ then $a_{ij}$ is by definition $\rho(e_{ij}) : \pi_j M \to \pi_i M$.

## 1. Dimension $(1,1)$

PROPOSITION 10.1. *The isomorphism classes of algebraically integral self-dual simple representations of $P_2$ with dimension vector $(1,1)$ correspond to triples of algebraic integers*

$$\left(\frac{1}{2} + r_1 i \ , \ \frac{1}{2} + r_2 i \ , \ r_3\right)$$

*with $r_1, r_2, r_3 \in \mathbb{R}$ and $r_3 \neq 0$.*

**1.1. Semisimple and Simple Representations.** If $\alpha = (1,1)$ then each $a_{ij}$ is a complex number and the representation space

$$R(P_2, (1,1)) = \bigoplus_{1 \leq i,j \leq 2} \text{Hom}(\mathbb{C}, \mathbb{C})$$

is 4-dimensional affine space. The coordinate ring is denoted

$$\mathbb{C}[X] = \mathbb{C}[X_{11}, X_{12}, X_{21}, X_{22}]$$

and the ring of invariant polynomials for the conjugation action of $\mathrm{GL}(1,1) \cong \mathbb{C}^\bullet \times \mathbb{C}^\bullet$ is generated by three algebraically independent (traces of) oriented cycles:
$$\mathbb{C}[X]^{\mathrm{GL}(1,1)} = \mathbb{C}[x_1, x_2, x_3]$$
with $x_1 = X_{11}$, $x_2 = X_{22}$ and $x_3 = X_{12}X_{21}$. The variety of semisimple dimension vector $(1,1)$ isomorphism classes of representations is therefore 3-dimensional affine space. A representation (28) is simple if and only if the complex number $a_{12}a_{21}$ is non-zero, so the open subvariety of simple representations is defined by $x_3 \neq 0$.

### 1.2. Self-Dual Representations.
A representation is self-dual if and only if its character $\chi : P_2 \to \mathbb{C}^-$ respects the involutions (lemma 8.8), i.e.
$$a_{11} = 1 - \overline{a_{11}}, \quad a_{22} = 1 - \overline{a_{22}} \quad \text{and} \quad a_{12}a_{21} = \overline{a_{12}a_{21}}.$$
Thus the isomorphism classes of self-dual semisimple representations correspond to the points in a 3-dimensional real affine space
$$\overline{\mathcal{M}}(P_2, (1,1)) = L \times L \times \mathbb{R} \tag{29}$$
where $L = \{\frac{1}{2} + bi \mid b \in \mathbb{R}\} \cong \mathbb{R}$.

Let us rewrite this computation from a dual point of view. The duality functor on $(P_2$–$\mathbb{C}^-)$-Proj induces an involution on $\mathbb{C}^-[x_1, x_2, x_3]$ given by
$$x_1 \mapsto 1 - x_1; \quad x_2 \mapsto 1 - x_2; \quad x_3 \mapsto x_3$$
(see chapter 6, section 3.4). The elements $i(x_1 - \frac{1}{2})$, $i(x_2 - \frac{1}{2})$ and $x_3$ are plainly involution invariant and the natural map
$$\mathbb{C}^- \otimes \mathbb{R}\left[i\left(x_1 - \frac{1}{2}\right), i\left(x_2 - \frac{1}{2}\right), x_3\right] \to \mathbb{C}^-[x_1, x_2, x_3]$$
is an isomorphism so the involution-invariant subring is
$$\mathbb{C}^-[X]^{\mathrm{GL}(1,1) \times \frac{\mathbb{Z}}{2\mathbb{Z}}} = \mathbb{R}\left[i\left(x_1 - \frac{1}{2}\right), i\left(x_2 - \frac{1}{2}\right), x_3\right].$$
Thus real points correspond to maximal ideals
$$\left(i\left(x_1 - \frac{1}{2}\right) - r_1, \; i\left(x_2 - \frac{1}{2}\right) - r_2, \; x_3 - r_3\right)$$
with $r_1, r_2, r_3 \in \mathbb{R}$. The maximal ideals of $\mathbb{C}^-[x_1, x_2, x_3]$ lying over these reals ideals correspond precisely to the elements of $L \times L \times \mathbb{R}$ which confirms (29) above.

### 1.3. Algebraically Integral Representations.
Recall from chapter 9 that a complex representation is said to be algebraically integral if it is a summand of a representation which is induced from an integral representation. By corollary 9.18, a semisimple complex representation of a quiver is algebraically integral if and only if all the traces of oriented cycles evaluate to algebraic integers. When $\alpha = (1,1)$ the trace of an oriented cycle is just some product of the generators $x_1$, $x_2$ and $x_3$, so the algebraically integral

## 2. Dimension $(2,1)$

representations correspond to triples of algebraic integers in $L \times L \times \mathbb{R}$. This completes the proof of proposition 10.1.

## 2. Dimension $(2,1)$

This section is devoted to the proof of the following proposition:

PROPOSITION 10.2. *The isomorphism classes of algebraically integral self-dual simple representations of $P_2$ with dimension vector $(2,1)$ correspond to quintuples of algebraic integers*

$$(30) \qquad \left(1 + 2r_1 i \ , \ r_2 + r_1 i \ , \ 2r_3 \ , \ r_3 + r_4 i \ , \ \frac{1}{2} + r_5 i\right)$$

*such that $r_1, \cdots, r_5 \in \mathbb{R}$ and $r_3^2 - 2r_1 r_4 r_3 + r_4^2 \neq 4r_2 r_3^2$.*

### 2.1. Semisimple Representations.
Fixing the dimension vector $\alpha = (2,1)$, the representation space $R(P_2, (2,1))$ is 9-dimensional and has coordinate ring $\mathbb{C}[X]$ where $X = \{X_{ij}\}_{1 \leq i,j \leq 3}$. Let $Y_{ij}$ denote the matrix of indeterminates corresponding to the linear map $a_{ij}$ so

$$\left(\begin{array}{c|c} Y_{11} & Y_{12} \\ \hline Y_{21} & Y_{22} \end{array}\right) = \left(\begin{array}{cc|c} X_{11} & X_{12} & X_{13} \\ X_{21} & X_{22} & X_{23} \\ \hline X_{31} & X_{32} & X_{33} \end{array}\right).$$

PROPOSITION 10.3. *The invariant ring $\mathbb{C}[X]^{\mathrm{GL}(2,1)}$ is generated by the following five polynomials:*

$$z_1 = \mathrm{Trace}(Y_{11}) = X_{11} + X_{22}$$
$$z_2 = \mathrm{Det}(Y_{11}) = X_{11} X_{22} - X_{12} X_{21}$$
$$z_3 = \mathrm{Trace}(Y_{12} Y_{21}) = X_{13} X_{31} + X_{23} X_{32}$$
$$z_4 = \mathrm{Trace}(Y_{11} Y_{12} Y_{21}) = \begin{pmatrix} X_{31} & X_{32} \end{pmatrix} \begin{pmatrix} X_{11} & X_{12} \\ X_{21} & X_{22} \end{pmatrix} \begin{pmatrix} X_{13} \\ X_{23} \end{pmatrix}$$
$$z_5 = \mathrm{Trace}(Y_{22}) = Y_{22} = X_{33}$$

*Moreover these polynomials are algebraically independent which implies that $\mathcal{M}(P_2, (2,1))$ is five-dimensional affine space.*

PROOF. It is easy to see that each $z_i$ is $\mathrm{GL}(\alpha)$-invariant, i.e.

$$\mathbb{C}[z_1, \cdots, z_5] \subset \mathbb{C}[X]^{\mathrm{GL}(\alpha)}.$$

To prove that $z_1, \cdots, z_5$ generate $\mathbb{C}[X]^{\mathrm{GL}(\alpha)}$ as a $\mathbb{C}$-algebra it suffices to show that the trace of every oriented cycle lies in $\mathbb{C}[z_1, \cdots, z_5]$. In fact we prove slightly more:

LEMMA 10.4. *The trace of every oriented cycle lies in $\mathbb{Z}[z_1, \cdots, z_5]$.*

PROOF. Let $Z$ be a non-trivial oriented cycle. $Y_{22}$ commutes with other oriented cycles at vertex 2 so we may assume that there are no occurrences of $Y_{22}$ in $Z$ and aim to show that $\mathrm{Trace}(Z) \in \mathbb{Z}[z_1, \cdots, z_4]$. Furthermore, we may assume that the cycle $Z$ begins and ends at vertex 1, if necessary by applying some equation of the form

$$\text{Trace}(Y_{21}Z'Y_{12}) = \text{Trace}(Z'Y_{12}Y_{21}).$$

Let $c_1 = Y_{11}$ and $c_2 = Y_{12}Y_{21}$ denote the two basic cycles at vertex 1. An arbitrary cycle $Z$ is a word in the alphabet $\{c_1, c_2\}$ and we have

$$z_1 = \text{Trace}(c_1); \quad z_2 = \text{Det}(c_1); \quad z_3 = \text{Trace}(c_2); \quad z_4 = \text{Trace}(c_1c_2).$$

Noting that $\det(c_2) = 0$, the cycle $Z$ may be expressed in terms of shorter cycles by means of the following Cayley-Hamilton identities:

(31) $$c_1^2 = \text{Trace}(c_1)c_1 - \text{Det}(c_1) = z_1c_1 - z_2$$

(32) $$c_2^2 = \text{Trace}(c_2)c_2 = z_3c_2$$

(33) $$c_1c_2c_1c_2 = \text{Trace}(c_1c_2)c_1c_2 = z_4c_1c_2.$$

Indeed, every sufficiently long word in $c_1$ and $c_2$ contains one of the three subwords $c_1^2$, $c_2^2$ or $c_1c_2c_1c_2$; the only words which do not contain one of these three are the following:

(34) $$c_1, \ c_2, \ c_1c_2, \ c_2c_1, \ c_1c_2c_1, \ c_2c_1c_2.$$

So $\text{Trace}(Z)$ is a linear combination (with coefficients in $\mathbb{Z}[z_1, \cdots, z_4]$) of the traces of these cycles (34). Now $\text{Trace}(c_1c_2c_1)$ and $\text{Trace}(c_2c_1c_2)$ each simplify further because

$$\text{Trace}(c_1c_2c_1) = \text{Trace}(c_1^2c_2) \quad \text{and} \quad \text{Trace}(c_2c_1c_2) = \text{Trace}(c_2^2c_1)$$

and $\text{Trace}(c_2c_1) = \text{Trace}(c_1c_2) = z_4$ so $\text{Trace}(Z)$ lies in $\mathbb{Z}[z_1, \cdots, z_4]$. □

To prove the last sentence of proposition 10.3, it suffices to note that by remark 6.6 the dimension of the variety $\mathcal{M}(P_2, (2,1))$ of isomorphism classes of semisimple representations is 5. □

### 2.2. Simple Representations.

PROPOSITION 10.5. *The isomorphism classes of semisimple representations which are not simple correspond to the points on the subvariety defined by*

(35) $$z_1z_3z_4 - z_4^2 = z_2z_3^2.$$

PROOF. If $M$ is a semisimple representation of dimension $(2,1)$ but is not simple then there must be a (simple) summand of dimension $(1,0)$; for if $M$ decomposes as a direct sum of representations of dimension $(0,1)$ and $(2,0)$ respectively then the latter representation must decompose further. We may therefore express $M$ as a direct sum of two representations whose dimensions are $(1,1)$ and $(1,0)$.

Our aim is to show that the image of the direct sum morphism

(36) $$\oplus : \mathcal{M}(P_2, (1,1)) \times \mathcal{M}(P_2, (1,0)) \to \mathcal{M}(P_2, (2,1)).$$

is the variety given by equation (35).

It follows from the definitions of the $z_i$ in proposition 10.3 that the morphism 36 is dual to the map of coordinate rings

$$\theta : \mathbb{C}[z_1, \cdots, z_5] \to \mathbb{C}[x_1, x_2, x_3, y]$$
$$z_1 \mapsto x_1 + y$$
$$z_2 \mapsto x_1 y$$
$$z_3 \mapsto x_3$$
$$z_4 \mapsto x_1 x_3$$
$$z_5 \mapsto x_2$$

where $\mathbb{C}[x_1, x_2, x_3]$ is the coordinate ring of $\mathcal{M}(P_2, (1,1))$ as in section 1.1 and $\mathbb{C}[y]$ denotes the coordinate ring of $\mathcal{M}(P_2, (1,0))$. We must show that the kernel of $\theta$ is the ideal generated by $p = z_1 z_3 z_4 - z_4^2 - z_2 z_3^2$.

It is straightforward to check that $p$ is contained in $\mathrm{Ker}(\theta)$. To prove the converse note that the restriction of $\oplus$ to $\mathcal{M}^s(P_2, (1,1)) \times \mathcal{M}(P_2, (0,1))$ is injective and so the image of $\oplus$ is four-dimensional and $\mathrm{Ker}(\theta)$ is a prime of height one. Therefore we need only check that $p$ is irreducible. Indeed, $p$ is linear in $z_1$ and the coefficient $z_3 z_4$ is coprime to $-z_4^2 - z_2 z_3^2$. This completes the proof of proposition 10.5. $\square$

**2.3. Self-Dual Representations.** As we discussed in chapter 6, section 3.4 the duality functor on $\mathcal{M}(P_2, (2,1))$ induces the involution

$$I(X_{ij}) = \begin{cases} -X_{ji} & \text{if } j \neq i \\ 1 - X_{ii} & \text{if } j = i \end{cases}$$

on $\mathbb{C}^-[X]$. It follows from the definition of $z_1, \cdots, z_5$ (in proposition 10.3) that the restriction of this involution to the $\mathrm{GL}(2,1)$ invariant ring $\mathbb{C}^-[X]^{\mathrm{GL}(2,1)} = \mathbb{C}^-[z_1, \cdots, z_5]$ is given by:

$$z_1 \mapsto 2 - z_1$$
$$z_2 \mapsto z_2 - z_1 + 1$$
$$z_3 \mapsto z_3$$
$$z_4 \mapsto -z_4 + z_3$$
$$z_5 \mapsto 1 - z_5.$$

The elements $i(z_1 - 1)$, $2z_2 - z_1$, $z_3$, $i(z_3 - 2z_4)$ and $i\left(z_5 - \frac{1}{2}\right)$ are fixed by the involution and the natural map

$$\mathbb{C}^- \otimes_\mathbb{R} \mathbb{R}\left[i(z_1 - 1),\ 2z_2 - z_1,\ z_3,\ i(z_3 - 2z_4),\ i\left(z_5 - \frac{1}{2}\right)\right] \to \mathbb{C}^-[z_1, \cdots, z_5]$$

is an isomorphism so by lemma 6.17

$$\mathbb{C}^-[X]^{\mathrm{GL}(2,1) \rtimes \frac{\mathbb{Z}}{2\mathbb{Z}}} = \mathbb{R}\left[i(z_1 - 1),\ 2z_2 - z_1,\ z_3,\ i(z_3 - 2z_4),\ i\left(z_5 - \frac{1}{2}\right)\right].$$

Thus there is one real maximal ideal and therefore one self-dual semisimple isomorphism class of representations for each quintuple

$$\left(1 + 2r_1 i \,,\; r_2 + r_1 i \,,\; 2r_3 \,,\; r_3 + r_4 i \,,\; \frac{1}{2} + r_5 i\right) \tag{37}$$

with $r_1, \cdots, r_5 \in \mathbb{R}$.

**2.4. Self-Dual Simple Representations.** In section 2.2 above we showed that the non-simple dimension $(2, 1)$ isomorphism classes of representations lie on the hypersurface defined by equation (35). We must compute the intersection of this hypersurface with the real variety of self-dual representations. Substituting the quintuple (37) for $(z_1, \cdots, z_5)$ in (35) and simplifying we obtain

$$r_3^2 - 2r_1 r_4 r_3 + r_4^2 = 4r_2 r_3^2 \tag{38}$$

Self-dual simple representations correspond to quintuples (37) which do not satisfy equation (38)

**2.5. Algebraically Integral Representations.**

LEMMA 10.6. *A semisimple dimension $(2, 1)$ complex representation $M$ is algebraically integral if and only if the invariants $z_1, \cdots, z_5$ take algebraic integer values at $M$.*

PROOF. Apply lemmas 9.18 and 10.4. □

This completes the proof of proposition 10.2.

CHAPTER 11

# Number Theory Invariants

In this chapter we use the theory of symmetric and hermitian forms over division algebras to obtain invariants which distinguish elements of finite order in $W^\epsilon(P_\mu\text{-}\mathbb{Q})$ and hence, by the following lemma, to distinguish such elements in $W^\epsilon(P_\mu\text{-}\mathbb{Z})$.

LEMMA 11.1. *Given any ring $R$ with involution, the natural map*

$$W^\epsilon(R\text{-}\mathbb{Z}) \to W^\epsilon(R\text{-}\mathbb{Q})$$

*is injective.*

PROOF. Suppose $(M, \rho, \phi) \in H^\epsilon(R\text{-}\mathbb{Z})$ is a non-singular $\epsilon$-hermitian form which represents the zero class in $W^\epsilon(R\text{-}\mathbb{Q})$. By lemma 5.10, the induced form $\mathbb{Q} \otimes_\mathbb{Z} (M, \rho, \phi)$ is metabolic with metabolizer $L \subset \mathbb{Q} \otimes M$ say. It follows that $L \cap M$ metabolizes $(M, \rho, \phi)$. □

We enhance lemma 11.1 in section 3 below.

Devissage followed by hermitian Morita equivalence decompose $W^\epsilon(P_\mu\text{-}\mathbb{Q})$ as in corollary 5.4

$$W^\epsilon(P_\mu\text{-}\mathbb{Q}) \cong \bigoplus_M W^1(\text{End}_{(P_\mu\text{-}\mathbb{Q})} M)$$

with one summand for each isomorphism class of $\epsilon$-self-dual simple representations $M$ of $P_\mu$ over $\mathbb{Q}$. Each endomorphism ring $\text{End}(M)$ is a finite $\mathbb{Q}$-dimensional division algebra with involution. Sections 1 and 2 of the present chapter are therefore a summary of results in the Witt theory of such division algebras; fortunately, complete Witt invariants are available. We do not provide proofs in these sections but the relevant theory can be found in the books of Albert [1], Lam [47] and Scharlau [93, Chapters 6,8,10] and in papers of Lewis [62, 63, 64]. Closely related $L$-theory computations have also been performed by I.Hambleton and I.Madsen [34].

There are five distinct classes of algebras with involution to be considered; in four of these a local-global principle applies. In chapter 12 below we prove that if $\mu \geq 2$ *every* finite $\mathbb{Q}$-dimensional division algebra with involution is the endomorphism ring of some simple (integral) representation of $P_\mu$ over $\mathbb{Q}$, so all the five classes are germane. Recall, however, that in the case $\mu = 1$ of knot theory, or in the case of a split $F_\mu$-link, one need only consider algebraic number fields with non-trivial involution, class 2a in theorem 11.5.

It is natural to ask whether a complete set of torsion invariants of $W^\epsilon(R\text{–}\mathbb{Q})$ can be defined by completing $\mathbb{Q}$ at finite primes, since all the torsion-free invariants can be obtained by completing $\mathbb{Q}$ at the (unique) real prime (see part b) of theorem 7.7). The answer is negative; since there is a class of division algebra with involution whose Witt group does not satisfy a local-global principle it follows from Morita equivalence that the 'Hasse-Minkowski' map

$$W^\epsilon(P_\mu\text{–}\mathbb{Q}) \to \prod_p W^\epsilon(P_\mu\text{–}\mathbb{Q}_p)$$

is not injective - in fact the kernel is isomorphic to $\left(\frac{\mathbb{Z}}{2\mathbb{Z}}\right)^{\oplus \infty}$. Nonetheless, the vast majority of invariants of $W^\epsilon(P_\mu\text{–}\mathbb{Q})$ can be defined locally.

## 1. Division Algebras over $\mathbb{Q}$

The basic structure theorem (Albert [1, p149]) is that every finite-dimensional division algebra (indeed every central simple algebra) over an algebraic number field is a cyclic algebra. A cyclic algebra $E$ over a field $K$ may be defined by the following construction (see [93, p316-320] for example):

Let $L/K$ be a finite Galois extension, say of degree $m$, with cyclic Galois group $\text{Gal}(L/K)$. Let $\sigma$ be a generator of $\text{Gal}(L/K)$ and let $a_0 \in K^\bullet$ be any non-zero element. The cyclic algebra $E = E(L/K, \sigma, a_0)$ is the $m$-dimensional $L$-vector space

$$E := L.1 \oplus L.e \oplus L.e^2 \oplus \cdots \oplus L.e^{m-1}$$

with multiplication defined by: $e^i.e^j := e^{i+j}$, $e^m := a_0 \in L.1$ and $ea = \sigma(a)e$ for all $a \in L$. The center of $E$ is $K$ and the dimension of $E$ over $K$ is $m^2$.

Let us now describe an important class of cyclic algebras, namely the quaternion algebras. Suppose $\alpha$ and $\beta$ are non-zero elements in $K$ and $\alpha$ is non-square. Let $L = K(\sqrt{\alpha})$, let $\sigma$ be the non-trivial Galois automorphism $\sqrt{\alpha} \mapsto -\sqrt{\alpha}$ and let $a_0 = \beta$. Then the cyclic algebra $E(L/K, \sigma, \beta)$ is the quaternion algebra:

$$E = (\alpha, \beta)_K = K\langle i, j \mid i^2 = \alpha; j^2 = \beta; ij = -ji \rangle$$
$$= K.1 \oplus K.i \oplus K.j \oplus K.k$$

with multiplication defined by $i^2 = \alpha$, $j^2 = \beta$, $k^2 = -\alpha\beta$ and $k = ij = -ji$. $E$ is a non-commutative four-dimensional $K$-algebra with $K$-basis $1$, $i$, $j$ and $ij$.

Of course, not every cyclic algebra is a division ring. In particular a quaternion algebra $(\alpha, \beta)$ fails to be a division ring if and only if the norm form $\langle 1, -\alpha, -\beta, \alpha\beta \rangle$ is isotropic, i.e. if and only if there is a non-trivial equation $0 = x_1^2 - \alpha x_2^2 - \beta x_3^2 + \alpha\beta x_4^2$ with $x_1, x_2, x_3, x_4 \in K$ (see Scharlau [93, p76]).

## 1. DIVISION ALGEBRAS OVER $\mathbb{Q}$

**1.1. Kinds of Involution.** One must distinguish two kinds of involution. Suppose $E$ is an algebra with center a field $K$ and let $I : E \to E^o$ be an involution. The restriction $I|_K : K \to K$ is an automorphism satisfying $I|_K^2 = \mathrm{id}_K$.

DEFINITION 11.2. The involution $I$ is said to be *of the first kind* if its restriction to the center is the identity automorphism $I|_K = \mathrm{id}_K$. Otherwise, $I$ is said to be *of the second kind*.

REMARK 11.3. If $M$ is an $\epsilon$-self-dual simple representation then the involution
$$\mathrm{End}(M) \to \mathrm{End}(M); \ f \mapsto b^{-1}f^*b$$
depends in general on a choice of hermitian (or skew-hermitian) form $b : M \to M^*$. However, distinct choices yield involutions which are conjugate and, in particular, are of the same kind. They are not, however, isomorphic in general.

In more detail, if $b,b' : M \to M^*$ are non-singular hermitian forms then we may write $b' = bc$ for some invertible $c \in \mathrm{End}(M)$ so
$$b'^{-1}f^*b' = (bc)^{-1}f^*(bc) = c^{-1}b^{-1}f^*bc$$
for all $f \in \mathrm{End}(M)$. Thus we may speak of simple self-dual representations of the first and second kind.

**1.2. Involutions of the First Kind.** The following theorem of Albert [**1**, p161] can also be found in Scharlau [**93**, p306,354]:

THEOREM 11.4. *i) A central simple algebra $A$ over a field $K$ admits an involution of the first kind if and only if $A \cong A^o$.*
*ii) If $K$ is an algebraic number field and $A$ is a central simple algebra with center $K$ such that $A \cong A^o$ then either $A = K$ or $A$ is a quaternion algebra.*

Henceforth we assume that $K$ is a number field. It follows quickly from theorem 11.4 that the simple algebras with involution fall into three classes as follows:

1a) $A = K$ is a number field with trivial involution;
1b) $A = K.1 \oplus K.i \oplus K.j \oplus K.k$ is a quaternion algebra with the 'standard' involution $a \mapsto \bar{a}$ defined by
$$\bar{i} = -i, \quad \bar{j} = -j, \quad , \bar{k} = -k$$
which fixes precisely $K$;
1c) $A = K.1 \oplus K.i \oplus K.j \oplus K.k$ is a quaternion algebra with 'non-standard' involution $a \mapsto \hat{a}$ defined by
$$\hat{i} = -i, \quad \hat{j} = j, \quad \hat{k} = k$$
which fixes a three-dimensional $K$-vector space.

Since $\hat{a} = i^{-1}\bar{a}i$ for all $a \in A$, multiplication by $i$ transforms an hermitian form over a quaternion algebra with non-standard involution $a \mapsto \hat{a}$ into a

skew-hermitian form over the same algebra with standard involution $a \mapsto \bar{a}$ so
$$W^1(A, \frown) \cong W^{-1}(A, ^-).$$
It is more difficult to define complete Witt invariants in case 1c) than in the other cases because the local-global principle fails. One requires a secondary invariant which is defined if all local invariants vanish - see section 2.3 below.

**1.3. Involutions of the Second Kind.** Suppose we are given an involution $^- : K \to K$ which fixes an index 2 subfield $k \subset K$ say. We distinguish just two classes of algebras with involutions of the second kind. Either

    2a) $A = K$ or
    2b) $A$ is a cyclic algebra over $K$.

There is no need to draw any distinction between hermitian and skew-hermitian forms, for if $K = k(\sqrt{\alpha})$ then $\overline{\sqrt{\alpha}} = -\sqrt{\alpha}$ and the transformation $\phi \mapsto \phi\alpha$ converts hermitian forms into skew-hermitian forms (and vice versa) so $W^1(K) \cong W^{-1}(K)$

## 2. Witt Invariants

The simplest Witt invariant of a symmetric or hermitian form is the rank $m$ modulo 2. Of course, if one or more signature invariants are defined then this rank invariant can often be deduced from the signatures. An exhaustive multi-signature invariant $\sigma$ has already been defined in earlier chapters. We will shortly recall the definitions of two further local invariants - the discriminant $\Delta$ and the Hasse-Witt invariant $c$. In the case of quaternionic algebras with non-standard involution, one requires an extra relative invariant such as the invariant $\theta$ which was introduced by D.Lewis.

Among many possible sources, the reader is referred to Landherr [49], Lewis [64], Milnor and Husemoller [70], O'Meara [75] and Scharlau [93] for a proof of the following theorem:

THEOREM 11.5. *The following table indicates a sufficient set of invariants to distinguish Witt classes of forms over each of the five classes of finite-dimensional division $\mathbb{Q}$-algebra:*

| Class | DivisionAlgebra | Involution | Invariants |
|---|---|---|---|
| 1a | Commutative | Trivial | $m$ (2), $\sigma$, $\Delta$, $c$ |
| 1b | Quaternionic | Standard | $m$ (2), $\sigma$ |
| 1c | Quaternionic | Non-standard | $m$ (2), $\sigma$, $\Delta$, $\theta$ |
| 2a | Commutative | Non-trivial | $m$ (2), $\sigma$, $\Delta$ |
| 2b | Cyclic | Second Kind | $m$ (2), $\sigma$, $\Delta$ |

Note that in the case 1c the invariant $\theta$ is defined only if all the other invariants vanish.

COROLLARY 11.6. *Let $R$ be any ring with involution. The subgroup $8W^\epsilon(R\text{--}\mathbb{Q}) \subset W^\epsilon(R\text{--}\mathbb{Q})$ is torsion-free.*

PROOF. By corollary 5.4 $W^\epsilon(R\text{–}\mathbb{Q})$ is isomorphic to a direct sum of Witt groups of division algebras $W^1(\text{End}(M))$. It follows from theorem 11.5, that the exponent of each summand $W^1(\text{End}(M))$ divides eight (compare caveats 11.7 and 11.8 below.) □

### 2.1. Discriminant $\Delta$.

1a. The *determinant* of a non-singular symmetric form $\phi : K^m \to (K^m)^*$ over any field $K$ (with trivial involution) is the determinant of the $m \times m$ matrix $A$ representing $\phi$ with respect to some choice of basis. Changing the basis one has $\det(PAP^t) = \det(A)\det(P)^2$ so $\det(\phi)$ is a well-defined element of $\frac{K^\bullet}{(K^\bullet)^2}$ where $(K^\bullet)^2 = \{x^2 \mid x \in K^\bullet\}$.

The *discriminant* of $\phi$ is, by definition,

$$\Delta(\phi) := (-1)^{\frac{m(m-1)}{2}} \det(\phi) \in \frac{K^\bullet}{(K^\bullet)^2}.$$

The discriminant vanishes on hyperbolic forms and is therefore a well-defined Witt invariant:

$$\Delta : W^1(K) \to \frac{K^\bullet}{(K^\bullet)^2} .$$

CAVEAT 11.7. $\Delta$ is not a group homomorphism. On the other hand, the rank modulo 2 and the discriminant together with a group homomorphism

$$W^1(K) \to \frac{\mathbb{Z}}{2\mathbb{Z}} \times \frac{K^\bullet}{(K^\bullet)^2}.$$

where the group operation on the right-hand side is '$\frac{\mathbb{Z}}{2\mathbb{Z}}$-graded':

$$(0, a) + (0, b) = (0, ab)$$
$$(0, a) + (1, b) = (1, ab)$$
$$(1, a) + (1, b) = (0, -ab) .$$

1b. Suppose next that $K$ is a field with a non-trivial involution $a \mapsto \bar{a}$ which fixes a subfield $k$ of index 2. The discriminant of a hermitian form $\phi$ over $K$ is defined as above, but the value group is slightly different:

$$\Delta : W^1(K) \to \frac{k^\bullet}{K^\bullet \overline{K^\bullet}}$$

$$[\phi] \mapsto (-1)^{\frac{m(m-1)}{2}} \det(\phi).$$

To explain the notation, $K^\bullet \overline{K^\bullet} = \{x\bar{x} \mid x \in K^\bullet\}$.

1c and 2b. Finally, if $E$ is a division ring of dimension $d^2$ over its center $K$ and $\phi : E^m \to (E^m)^*$ is an hermitian form then the determinant of $\phi$ is by definition the reduced norm of a matrix representing $\phi$. It can be computed as follows: Choose a Galois extension $L$ of $K$ which splits $E$, i.e. such that $L \otimes_K E \cong M_d(L)$. Then $\phi$ induces a form $\phi_{L\otimes E}$ which is represented by a

$dn \times dn$ matrix over $L$ and whose determinant is $\det(\phi)$. To obtain a Witt invariant one defines the discriminant

$$\Delta(\phi) = (-1)^{\frac{m(m-1)}{2}d} \det(\phi)$$

which takes values in $\frac{K^\bullet}{(K^\bullet)^2}$ if the involution is of the first kind or in $\frac{k^\bullet}{K^\bullet K^\bullet}$ if the involution is of the second kind. Note that if $d$ is even, such as in the case 1c, then we have $\Delta(\phi) = \det(\phi)$.

## 2.2. Hasse-Witt Invariant $c$.
Suppose $K$ is an algebraic number field with trivial involution and let $\Psi$ denote the set of primes. For each prime $\mathfrak{p} \in \Psi$ and each pair of units $a, b \in K^\bullet$ there is defined a Hilbert symbol

$$(a, b)_\mathfrak{p} = \begin{cases} 1 & \text{if there exist } x, y \in K_\mathfrak{p} \text{ such that } ax^2 + by^2 = 1 \\ -1 & \text{otherwise} \end{cases}$$

Equivalently, $(a, b)_\mathfrak{p} = 1$ if and only if the quaternion algebra $(a, b)_{K_\mathfrak{p}}$ is isomorphic to the matrix algebra $M_2(K_\mathfrak{p})$. The number of primes at which $(a, b)_\mathfrak{p} = -1$ is finite (and, according to Hilbert reciprocity, even) so, considering all primes together, we may write

$$(a, b) \in \bigoplus_\Psi \{+1, -1\}.$$

Given a non-singular symmetric form $\phi = \langle a_1, a_2, \cdots, a_m \rangle$ over $K$ let

$$s(\phi) = \prod_{i<j}(a_i, a_j) \in \bigoplus_\Psi \{+1, -1\}.$$

It turns out that $s$ is independent of the diagonalization. Some adjustment is needed to obtain a well-defined Witt invariant $c$ [93, p81]:

$$c(\phi) := \begin{cases} s(\phi) & \text{if } m \cong 1, 2 \pmod{8} \\ (-1, -\det(\phi))s(\phi) & \text{if } m \cong 3, 4 \pmod{8} \\ (-1, -1)s(\phi) & \text{if } m \cong 5, 6 \pmod{8} \\ (-1, \det(\phi))s(\phi) & \text{if } m \cong 7, 8 \pmod{8} \end{cases}.$$

CAVEAT 11.8. The Hasse-Witt invariant $c : W^1(K) \to \bigoplus_\Psi \{+1, -1\}$ is not a homomorphism. However, the rank modulo 2, the discriminant and the Hasse-Witt invariant together give a homomorphism

$$W^1(K) \to \frac{\mathbb{Z}}{2\mathbb{Z}} \times \frac{K^\bullet}{\{a^2 \mid a \in K^\bullet\}} \times \left(\bigoplus_\Psi \{+1, -1\}\right)$$

where the group law on the right hand side is given by

$$(0, d, c) + (0, d', c') = (0, dd', (d, d')cc')$$
$$(0, d, c) + (1, d', c') = (1, dd', (d, -d')cc')$$
$$(1, d, c) + (0, d', c') = (1, dd', (-d, d')cc')$$
$$(1, d, c) + (1, d', c') = (0, -dd', (d, d')cc').$$

**2.3. Local-Global Principle.** Naturality of devissage and hermitian Morita equivalence (remarks 5.11 and 4.10) give rise to a commutative diagram:

$$\begin{array}{ccccc}
W^\epsilon(R\text{-}\mathbb{Q}) & \longrightarrow & & & \prod_p W^\epsilon(R\text{-}\mathbb{Q}_p) \\
\uparrow & & & & \uparrow \\
\bigoplus_M W^\epsilon_M(R\text{-}\mathbb{Q}) & \longrightarrow & \bigoplus_M \prod_p W^\epsilon_{\mathbb{Q}_p \otimes M}(R\text{-}\mathbb{Q}_p) & \rightarrowtail & \prod_p \bigoplus_M W^\epsilon_{\mathbb{Q}_p \otimes_\mathbb{Q} M}(R\text{-}\mathbb{Q}_p) \\
\cong \updownarrow & & \cong \updownarrow & & \cong \updownarrow \\
\bigoplus_M W^1(E_M) & \longrightarrow & \bigoplus_M \prod_p W^1(\mathbb{Q}_p \otimes_\mathbb{Q} E_M) & \rightarrowtail & \prod_p \bigoplus_M W^1(\mathbb{Q}_p \otimes_\mathbb{Q} E_M)
\end{array}$$

where $E_M = \text{End}_{(R\text{-}\mathbb{Q})} M$. All the sums are indexed by the isomorphism classes of $\epsilon$-self-dual simple representations $M$ of $R$ over $\mathbb{Q}$.

When $E_M$ is not a quaternion algebra with non-standard involution (i.e. not in class 1c above), a local global principle applies. In other words, the natural map

$$(39) \qquad W^1(E_M) \to \prod_p W^1(\mathbb{Q}_p \otimes_\mathbb{Q} E_M)$$

is injective.

**2.4. Lewis $\theta$-invariant.** Suppose $E = K(a,b) = K\langle i, j \mid i^2 = a; j^2 = b; ij = -ji\rangle$ is a quaternion division algebra with non-standard involution $\hat{i} = -i, \hat{j} = j$. The rank modulo 2, the signatures and the discriminant are together a complete set of local Witt invariants; that is to say, if two Witt classes cannot be distinguished by these three invariants then they have the same image under the natural map (39).

One further relative invariant is needed to distinguish elements in the kernel of (39). The first such invariant was constructed by Bartels [3, 4] using Galois cohomology. We shall describe instead the more elementary invariant $\theta$ which is due to Lewis [64].

Let $L = K(\sqrt{a})$ and let $W^1(L)$ and $W^1(L, \bar{\ })$ denote the Witt groups of $L$ with trivial involution and involution $\sqrt{a} \mapsto -\sqrt{a}$ fixing $K$ respectively.

The definition of $\theta$ involves the following commutative diagram:

$$\begin{array}{ccccccccc}
0 & \longrightarrow & W^1(E,\bar{\ }) & \longrightarrow & W^1(L,\bar{\ }) & \stackrel{\iota}{\longrightarrow} & W^1(E,\hat{\ }) & \longrightarrow & W^1(L)\cdots \\
& & \downarrow & & \downarrow & & \downarrow \Lambda & & \downarrow \Lambda' \\
0 & \longrightarrow & \prod W^1(E_\mathfrak{p},\bar{\ }) & \longrightarrow & \prod W^1(L_\mathfrak{p},\bar{\ }) & \longrightarrow & \prod W^1(E_\mathfrak{p},\hat{\ }) & \longrightarrow & \prod W^1(L_\mathfrak{p})\cdots
\end{array}$$

in which the horizontal sequences are exact (Lewis [**61, 64, 63**]) and the products are indexed by the primes $\mathfrak{p}$ of $K$. The map $\iota$ is induced by the inclusion of $L = K(\sqrt{a}) \hookrightarrow E$; $\sqrt{a} \mapsto i$. Exactness of the upper row and the injectivity of $\Lambda'$ is that $\mathrm{Ker}(\Lambda) \subset \mathrm{Im}(\iota)$. Given any $\phi \in \mathrm{Ker}(\Lambda)$ one can choose $\psi \in W^1(L,\bar{\ })$ such that $\iota(\psi) = \phi$ and let $d$ be the discriminant

$$d = \Delta(\psi) \in \frac{K^\bullet}{L^\bullet \overline{L}^\bullet}.$$

There is now a well-defined injection

$$\theta : \mathrm{Ker}(\Lambda) \to \{+1,-1\}^S/\sim$$
$$[\phi] \mapsto \{(d,a)_\mathfrak{p}\}_{\mathfrak{p} \in S}$$

where $S$ is the set of primes at which $E_\mathfrak{p}$ is a division algebra and where $\sim$ identifies each element $(\epsilon_1, \cdots, \epsilon_{|S|})$ with its antipode $(-\epsilon_1, \cdots, -\epsilon_{|S|})$.

## 3. Localization Exact Sequence

Although we shall not attempt to extricate the subgroup $W^\epsilon(P_\mu\text{–}\mathbb{Z})$ from $W^\epsilon(P_\mu\text{–}\mathbb{Q})$ in any detail, a first step in that direction is the following localization exact sequence which makes sense for any ring $R$ with involution:

(40) $\qquad 0 \to W^\epsilon(R\text{–}\mathbb{Z}) \to W^\epsilon_\mathbb{Z}(R\text{–}\mathbb{Q}) \to W^\epsilon(R\text{–}\mathbb{Q}/\mathbb{Z}) \to \cdots$

In this section we define and analyse the groups $W^\epsilon_\mathbb{Z}(R\text{–}\mathbb{Q})$ and $W^\epsilon(R\text{–}\mathbb{Q}/\mathbb{Z})$. We omit to prove that (40) is exact since the argument is quite standard - e.g. Stoltzfus [**97**, pp16-19] or Neumann [**73**, Theorem 6.5]).

DEFINITION 11.9. *If $R$ is any ring with involution, let $(R\text{–}\mathbb{Q})_\mathbb{Z}$-Proj denote the full subcategory of $(R\text{–}\mathbb{Q})$-Proj containing precisely those rational representations $(M,\rho)$ which are induced up from integral representations, i.e. $(M,\rho) \cong \mathbb{Q} \otimes_\mathbb{Z} (M_0,\rho_0)$.*

DEFINITION 11.10. *Let $W^\epsilon_\mathbb{Z}(R\text{–}\mathbb{Q})$ denote the Witt group of $(R\text{–}\mathbb{Q})_\mathbb{Z}$-Proj.*

LEMMA 11.11. *Suppose we are given a short exact sequence in $(R\text{–}\mathbb{Q})$-Proj*

$$0 \to M \to M' \to M'' \to 0.$$

*If $M' \in (R\text{–}\mathbb{Q})_\mathbb{Z}$-Proj then $M$ and $M''$ are also in $(R\text{–}\mathbb{Q})_\mathbb{Z}$-Proj. It follows, in particular, that $(R\text{–}\mathbb{Q})_\mathbb{Z}$-Proj is an abelian category.*

PROOF. If $M' \cong \mathbb{Q} \otimes_\mathbb{Z} M'_0$ then

$$M \cong \mathbb{Q} \otimes_\mathbb{Z} (M'_0 \cap M) \quad \text{and} \quad M'' \cong \mathbb{Q} \otimes_\mathbb{Z} \frac{M'_0}{M \cap M'_0}. \qquad \square$$

## 3. LOCALIZATION EXACT SEQUENCE

PROPOSITION 11.12. *There is an isomorphism*

$$W^\epsilon_\mathbb{Z}(R\text{--}\mathbb{Q}) \cong \bigoplus_M W^1(\mathrm{End}(M)) \tag{41}$$

*with one direct summand for each simple $\epsilon$-self-dual $M \in (R\text{--}\mathbb{Q})_\mathbb{Z}$-Proj. In particular $W^\epsilon_\mathbb{Z}(R\text{--}\mathbb{Q})$ is a direct summand of $W^\epsilon_\mathbb{Z}(R\text{--}\mathbb{Q})$.*

PROOF. The first sentence is a consequence of theorems 5.3 and 4.7. It follows from lemma 11.11 that simple objects in $(R\text{--}\mathbb{Q})_\mathbb{Z}$-Proj are simple in $(R\text{--}\mathbb{Q})$-Proj which implies the second sentence of the proposition. □

Proposition 9.14 showed that a semisimple rational representation $M \in (R\text{--}\mathbb{Q})$-Proj lies in $(R\text{--}\mathbb{Q})_\mathbb{Z}$-Proj if and only if the character $\chi_M$ takes values in $\mathbb{Z}$. Let us now give a criterion which applies also to non-semisimple representations.

LEMMA 11.13. *Suppose $(M, \rho) \in (R\text{--}\mathbb{Q})$-Proj. Then the following are equivalent:*

(1) $M \in (R\text{--}\mathbb{Q})_\mathbb{Z}$-Proj.
(2) *$M$ is algebraically integral. In other words there exists a rational representation $M' \in (R\text{--}\mathbb{Q})$-Proj and an integral representation $N_0 \in (R\text{--}\mathbb{Z})$-Proj such that $M \oplus M' \cong \mathbb{Q} \otimes_\mathbb{Z} M_0$.*
(3) *The subring $\rho(R) \subset \mathrm{End}_\mathbb{Q} M$ is finitely generated as a $\mathbb{Z}$-module.*

PROOF. $1 \Rightarrow 2$: Immediate.
$2 \Rightarrow 1$: Suppose $M \oplus M' \simeq \mathbb{Q} \otimes N_0$ for some $M' \in (R\text{--}\mathbb{Q})$-Proj and some $N_0 \in (R\text{--}\mathbb{Z})$-Proj. The intersection $M_0 = N_0 \cap M$ is invariant under the action of $R$ and satisfies $M \cong \mathbb{Q} \otimes_\mathbb{Z} M_0$.
$1 \Rightarrow 3$: The representation $\rho$ factors through a canonical inclusion

$$\mathrm{End}_\mathbb{Z} M_0 \hookrightarrow \mathrm{End}_\mathbb{Q} M.$$

Now $\rho(R)$ is contained in the finitely generated free $\mathbb{Z}$-module $\mathrm{End}_\mathbb{Z} M_0$. Since $\mathbb{Z}$ is Noetherian, $\rho(R)$ is finitely generated.
$3 \Rightarrow 1$: If $\rho(R)$ is finitely generated then $\rho(R)x$ is a finitely generated sub-$\mathbb{Z}$-module for every $x \in M$. If $x_1, \cdots, x_m$ is a basis for $M$ over $\mathbb{Q}$ then we may set $M_0 = \sum_i \rho(R) x_i$. □

We turn next to the group $W^\epsilon(R\text{--}\mathbb{Q}/\mathbb{Z})$.

DEFINITION 11.14. We denote by $(R\text{--}\mathbb{Q}/\mathbb{Z})$-Proj the abelian category of representations $(M, \rho)$ where $M$ is a finite abelian group and $\rho : R \to \mathrm{End}_\mathbb{Z} M$ is a ring homomorphism.

The dual module $M^\wedge = \mathrm{Hom}_\mathbb{Z}(M, \mathbb{Q}/\mathbb{Z})$ admits the usual $R$-action

$$\rho^\wedge(r)(\theta) = (m \mapsto \theta(\bar{r}m)) \quad \text{for all } \theta \in M^\wedge, r \in R \text{ and } m \in M.$$

so we have defined a duality functor on $(R\text{--}\mathbb{Q}/\mathbb{Z})$-Proj.

DEFINITION 11.15. Let $W^\epsilon(R\text{--}\mathbb{Q}/\mathbb{Z})$ denote the Witt group of the hermitian category $(R\text{--}\mathbb{Q}/\mathbb{Z})$-Proj.

## 11. NUMBER THEORY INVARIANTS

The machinery described in chapters 4 and 5 is general enough to compute this Witt group since $(R\text{–}\mathbb{Q}/\mathbb{Z})$-Proj is an abelian category. By theorems 5.3 and 4.7 $W^\epsilon(R\text{–}\mathbb{Q}/\mathbb{Z})$ is a direct sum of Witt groups of finite fields. These are easy to compute since the group of units of a finite field is cyclic:

LEMMA 11.16. *If $K$ is a finite field with trivial involution then*

$$W^1(K) \cong \begin{cases} \frac{\mathbb{Z}}{2\mathbb{Z}} \oplus \frac{\mathbb{Z}}{2\mathbb{Z}} & \text{if } |K| \equiv 1 \pmod 4 \\ \frac{\mathbb{Z}}{4\mathbb{Z}} & \text{if } |K| \equiv 3 \pmod 4 \\ \frac{\mathbb{Z}}{2\mathbb{Z}} & \text{if } \operatorname{char}(K) = 2. \end{cases}$$

*If the involution is non-trivial then*

$$W^1(K) \cong \frac{\mathbb{Z}}{2\mathbb{Z}}.$$

PROOF. See for example Scharlau [**93**, p40] and Milnor and Husemoller [**70**, p117]. □

Suppose $p$ is an odd prime. It is easy to construct examples of self-dual objects in $(P_\mu\text{–}\mathbb{Q}/\mathbb{Z})$-Proj whose endomorphism ring has $p$ elements. For example, let $s$ act on $\frac{\mathbb{Z}}{p\mathbb{Z}}$ as multiplication by $\frac{p+1}{2}$. We obtain:

PROPOSITION 11.17. *There is an isomorphism*

$$W^\epsilon(P_\mu\text{–}\mathbb{Q}/\mathbb{Z}) \cong \bigoplus_M W^1(\operatorname{End}(M)) \cong \left(\frac{\mathbb{Z}}{4\mathbb{Z}}\right)^{\oplus \infty} \oplus \left(\frac{\mathbb{Z}}{2\mathbb{Z}}\right)^{\oplus \infty}.$$

*There is one summand $W^1(\operatorname{End}(M))$ for each isomorphism class of $\epsilon$-self-dual simple representations $M \in (P_\mu\text{–}\mathbb{Q}/\mathbb{Z})$-Proj.*

# CHAPTER 12

# All Division Algebras Occur

We prove here that every finite-dimensional division algebra with involution is the endomorphism ring over $(P_\mu\text{–}\mathbb{Q})$ of some simple $\epsilon$-self-dual integral representation of $P_\mu$. Consequently, all five of the classes of division algebras highlighted in chapter 11 arise in the computation of boundary link cobordism (whereas only the class 2a of number fields with non-trivial involution occurs when $\mu = 1$). We conclude the chapter with a proof of theorem A.

Recall that the quiver path ring $P_\mu$ has presentation

$$\mathbb{Z}\left\langle s, \pi_1, \cdots, \pi_\mu \,\middle|\, \sum_{i=1}^{\mu} \pi_i = 1, \pi_i^2 = \pi_i, \pi_i\pi_j = 0 \text{ for } 1 \leq i, j \leq \mu \right\rangle.$$

and involution $s \mapsto 1 - s$, $\pi_i \mapsto \pi_i$ for $1 \leq i \leq \mu$. The endomorphism ring $\text{End}_{(P_\mu\text{–}\mathbb{Q})} M$ of a rational representation $(M, \rho)$ is
(42)
$$\left\{ (\alpha_1, \cdots, \alpha_\mu) \in \bigoplus_{i=1}^{\mu} \text{End}_\mathbb{Q}(\pi_i M) \,\middle|\, \alpha_i \rho(s)_{ij} = \rho(s)_{ij}\alpha_j \text{ for } 1 \leq i, j \leq \mu \right\}$$

where $\rho(s)_{ij}$ is the composite $\pi_j M \hookrightarrow M \xrightarrow{\rho(s)} M \twoheadrightarrow \pi_i M$.

PROPOSITION 12.1. *Suppose $E$ is a finite-dimensional division algebra over $\mathbb{Q}$ and let $\mu \geq 2$. There exists an integral representation $M_0$ of $P_\mu$, finitely generated and free over $\mathbb{Z}$, such that $M = \mathbb{Q} \otimes_\mathbb{Z} M_0$ is simple and $\text{End}_{(P_\mu\text{–}\mathbb{Q})}(M)$ is isomorphic to $E$.*

Recall that the opposite ring $E^o$ is identical to $E$ as an additive group but multiplication is reversed.

PROPOSITION 12.2. *Suppose $E$ is a finite-dimensional division algebra over $\mathbb{Q}$ with involution $I : E \to E^o$. Let $\epsilon = +1$ or $-1$ and let $\mu \geq 2$. There exists an $\epsilon$-self-dual integral representation $M_0$ of $P_\mu$ and an $\epsilon$-hermitian form $b_0 : M_0 \to M_0^*$ such that*
*i) $M_0$ is finitely generated and free over $\mathbb{Z}$.*
*ii) $M = \mathbb{Q} \otimes_\mathbb{Z} M_0$ is simple.*
*iii) There is an isomorphism*

$$(\text{End}_{(P_\mu\text{–}\mathbb{Q})} M \,,\, \beta \mapsto b^{-1}\beta^* b) \cong (E, I)$$

*of algebras with involution. Here, $b : M \to M^*$ is the isomorphism induced by $b_0$.*

The idea in the proof of proposition 12.1 will be to express $E$ as an endomorphism ring via the identity

(43) $$E \cong \operatorname{End}_{E^o} E^o$$

and let $P_\mu$ mimic the action of $E^o$ on $E^o$. Our proof of proposition 12.2 uses, in addition, the observation that for every involution

$$I : E = \operatorname{End}_{E^o} E^o \to \operatorname{End}_{E^o} E^{o*} = E^o$$

there exists $\delta : E^{o*} \to E^{o*}$ such that $I(x) = \delta^{-1} x^* \delta$ for all $x \in \operatorname{End}_{E^o} E^o$. Such $\delta$ exists because left vector spaces of equal dimension over $E^o$ are isomorphic.

## 1. Proof of Proposition 12.1

Let $l : E \to \operatorname{End}_{\mathbb{Q}} E$ and $r : E^o \to \operatorname{End}_{\mathbb{Q}} E$ denote the regular representations 'multiplication on the left' and 'multiplication on the right':

$$l(x)(y) = xy \quad \text{and} \quad r(x)(y) = yx \quad \text{for all } x, y \in E.$$

Note that $l(E) = Z_{\operatorname{End}_{\mathbb{Q}} E}(r(E^o))$ and $r(E^o) = Z_{\operatorname{End}_{\mathbb{Q}} E}(l(E))$ where

$$Z_{\operatorname{End}_{\mathbb{Q}} E}(S) = \{x \in \operatorname{End}_{\mathbb{Q}} E \mid xs = sx \text{ for all } s \in S\}$$

denotes the commutator of a subset $S \subset \operatorname{End}_{\mathbb{Q}} E$.

Let us consider first the case $\mu = 2$. By theorem I.1 of appendix I, $E$ is generated as a $\mathbb{Q}$-algebra by two elements $x_1$, $x_2$ say. We define a representation $M$ over $\mathbb{Q}$ as follows: Let $\pi_1 M \cong \pi_2 M \cong E$ as $\mathbb{Q}$-vector spaces and let $s \in P_2$ act via the matrix

(44) $$\left( \begin{array}{c|c} r(x_1) & 1 \\ \hline 1 & r(x_2) \end{array} \right).$$

By equation (42) we have

$$\operatorname{End}_{P_2}(M) = \{(\alpha, \alpha) \in \operatorname{End}_{\mathbb{Q}} E \times \operatorname{End}_{\mathbb{Q}} E \mid \alpha r(x_i) = r(x_i) \alpha \ (i = 1, 2)\}$$
$$\cong \{(\alpha, \alpha) \in \operatorname{End}_{\mathbb{Q}} E \times \operatorname{End}_{\mathbb{Q}} E \mid \alpha \in l(E)\}$$
$$\cong E.$$

Multiplying the generators $x_1$, $x_2$ by an integer if necessary we can ensure that $M \cong \mathbb{Q} \otimes M_0$ where $M_0$ is an integral representation, finitely generated over $\mathbb{Z}$.

We must also check that $M$ is a simple representation. Indeed, if $M'$ is a subrepresentation of $M$ then the action (44) of $s$ implies that $\pi_1 M' \cong \pi_2 M'$ and moreover that $\pi_1 M'$ is a sub-$E^o$-module of $E$. Thus $M' = 0$ or $M$ as required.

The construction extends easily to all $\mu \geq 2$. Let $\pi_i M = E$ for $i = 1, \cdots, \mu$ and define $\rho(s)_{ij} : \pi_j M \to \pi_i M$ by

$$\rho(s)_{ij} = \begin{cases} r(x_i) & \text{if } i = j \\ 1 & \text{if } i \neq j \end{cases}$$

where $x_1, x_2, \cdots, x_\mu$ is any set of generators for $E^o$ over $\mathbb{Q}$. This completes the proof of proposition 12.1.

REMARK 12.3. This proof yields infinitely many representations $M = \mathbb{Q} \otimes M_0$ with endomorphism ring $E$, no two of which are isomorphic. Indeed, one may add any integer $a \in \mathbb{Z}$ to either of the generators $x_1, x_2$ and continue to construct $M$ as before. Distinct choices of $a$ yield representations with distinct characters, so no two are isomorphic by corollary 8.6.

## 2. Proof of Proposition 12.2

**2.1. Construction of $M$ and $b$.** In outline, we begin with a representation $M$ constructed as in the previous section which is *not* self-dual but which satisfies $\mathrm{End}_\mathbb{Q} M \cong E$. The endomorphism ring of the direct sum $M \oplus M^*$ is $E \times E^o$ and the form $\begin{pmatrix} 0 & 1 \\ \epsilon & 0 \end{pmatrix} : M \oplus M^* \to (M \oplus M^*)^*$ induces the transposition involution on $E \times E^o$:

$$(45) \quad \overline{\begin{pmatrix} \beta & 0 \\ 0 & \beta' \end{pmatrix}} = \begin{pmatrix} 0 & 1 \\ \epsilon & 0 \end{pmatrix}^{-1} \begin{pmatrix} \beta & 0 \\ 0 & \beta' \end{pmatrix}^* \begin{pmatrix} 0 & 1 \\ \epsilon & 0 \end{pmatrix} = \begin{pmatrix} \beta' & 0 \\ 0 & \beta \end{pmatrix}.$$

We adjust the action of $P_\mu$ in such a way that the endomorphism ring is confined precisely to the graph of the involution

$$\{(\beta, I(\beta)) \in E \times E^o \mid \beta \in E\}.$$

To explain the construction in more detail, let us assume $\mu = 2$ and let $x_1, x_2$ be generators for $E$ over $\mathbb{Q}$ as before. We write

$$\pi_1 M \cong \pi_2 M \cong E \oplus E^*$$

with $E^* = \mathrm{Hom}_\mathbb{Q}(E, \mathbb{Q})$ and let $s$ act on $M$ via the matrix

$$(46) \quad \left( \begin{array}{cc|cc} r(x_1) & 0 & 1 & \gamma \\ 0 & 1 - r(x_1)^* & \delta & -1 \\ \hline 1 & -\epsilon\gamma^* & r(x_2) & 0 \\ -\epsilon\delta^* & -1 & 0 & 1 - r(x_2)^* \end{array} \right).$$

A very mild constraint on the choice of generators $x_1$ and $x_2$ will be imposed in the proof of lemma 12.4 and the maps $\gamma$ and $\delta$ will be defined below. We define an $\epsilon$-hermitian form

$$b = \left( \begin{array}{cc|cc} 0 & 1 & 0 & 0 \\ \epsilon & 0 & 0 & 0 \\ \hline 0 & 0 & 0 & 1 \\ 0 & 0 & \epsilon & 0 \end{array} \right).$$

It is easy to check that $bs = (1-s)^*b$, so $b : M \to M^*$ is a homomorphism of representations.

## 2.2. Calculation of $\text{End}_{(P_2-\mathbb{Q})} M$.

LEMMA 12.4. *The generators $x_1$ and $x_2$ can be chosen such that every endomorphism $\alpha \in \text{End}_{(P_2-\mathbb{Q})} M$ can be written*

$$\alpha = \begin{pmatrix} \beta_1 & 0 & 0 & 0 \\ 0 & \beta_1' & 0 & 0 \\ \hline 0 & 0 & \beta_2 & 0 \\ 0 & 0 & 0 & \beta_2' \end{pmatrix}.$$

*In other words*

$$\text{End}_{P_2}(M) \subset \text{End}_{\mathbb{Q}}(E) \times \text{End}_{\mathbb{Q}}(E^*) \times \text{End}_{\mathbb{Q}}(E) \times \text{End}_{\mathbb{Q}}(E^*).$$

PROOF. Certainly every endomorphism of $M$ is a direct sum of an endomorphism of $\pi_1 M$ and an endomorphism of $\pi_2 M$. We may think of the component $\pi_1 M$ as a representation of a polynomial ring $\mathbb{Z}[s_{11}]$ where $s_{11}$ acts via $\begin{pmatrix} r(x_1) & 0 \\ 0 & 1-r(x_1)^* \end{pmatrix} \in \text{End}_{\mathbb{Q}}(\pi_1 M)$.

The subfield $L_1 = \mathbb{Q}(x_1) \subset E$ generated by $x_1$ is a simple representation of $\mathbb{Z}[s_{11}]$ in which the action of $s_{11}$ is multiplication by $x_1$. Moreover, $E$ can be expressed as a direct sum of copies of these simple representations $E = L_1^{\oplus d}$. Dually, $E^*$ is a direct sum of simple representations $L_1^*$ on which $s_{11}$ acts as $1 - r(x_1)^*$.

Multiplying $x_1$ by an integer if necessary, we can assume that the endomorphism $r(x_1) \in \text{End}_{\mathbb{Q}} E$ is *not* isomorphic to $1 - r(x_1)^* \in \text{End}_{\mathbb{Q}}(E^*)$, so that $L_1$ and $L_1^*$ are not isomorphic representations. It follows that $\text{Hom}_{\mathbb{Z}[s_{11}]}(L_1^{\oplus d}, L_1^{*\oplus d}) = 0$ whence

$$\text{End}_{\mathbb{Z}[s_{11}]}(\pi_1 M) = \text{End}_{\mathbb{Z}[s_{11}]}(E) \times \text{End}_{\mathbb{Z}[s_{11}]}(E^*).$$

The same argument applies to $\pi_2(M)$, completing the proof of lemma 12.4. □

Equation (42) imposes relations on the components $\beta_1$, $\beta_1'$, $\beta_2$ and $\beta_2'$ of an endomorphism $\beta \in \text{End}_{(P_2-\mathbb{Q})} M$. In particular,

$$\begin{pmatrix} \beta_1 & 0 \\ 0 & \beta_1' \end{pmatrix} \begin{pmatrix} 1 & \gamma \\ \delta & -1 \end{pmatrix} = \begin{pmatrix} 1 & \gamma \\ \delta & -1 \end{pmatrix} \begin{pmatrix} \beta_2 & 0 \\ 0 & \beta_2' \end{pmatrix}$$

so $\beta_1 = \beta_2$ and $\beta_1' = \beta_2'$. If we write $\beta = \beta_1 = \beta_2$ and $\beta' = \beta_1' = \beta_2'$ then the relations imposed in (42) can be summarized as follows:

(47) $\quad\quad\quad\quad\quad \beta r(x_i) = r(x_i)\beta \quad \text{for } i = 1, 2;$

(48) $\quad\quad\quad\quad\quad \beta' r(x_i)^* = r(x_i)^* \beta' \quad \text{for } i = 1, 2;$

(49) $\quad\quad\quad\quad\quad \beta\gamma = \gamma\beta'; \quad\quad\quad\quad\quad \beta\gamma^* = \gamma^*\beta';$

(50) $\quad\quad\quad\quad\quad \beta'\delta = \delta\beta; \quad\quad\quad\quad\quad \beta'\delta^* = \delta^*\beta.$

Equation (47) is equivalent to the statement that $\beta \in \text{End}_{\mathbb{Q}} E$ is left multiplication by some element $x \in E$. Similarly, equation (48) is equivalent to the statement that $\beta' \in \mathrm{End}_{\mathbb{Q}}(E^*)$ is dual to left multiplication by some $x' \in E$:

$$\beta = l(x); \qquad \beta' = l(x')^*.$$

The adjoint involution $\alpha \mapsto b^{-1}\alpha^* b$ is the transposition $(l(x), l(x')^*) \mapsto (l(x'), l(x)^*)$ by equation (45). We wish to choose $\gamma$ and $\delta$ in such a way that equations (49) and (50) impose precisely the condition $x' = I(x)$.

If $\delta$ is invertible and satisfies $\delta l(x) = l(I(x))^* \delta$ for all $x \in E$ then the equations (50) impose the equivalent relations $x' = I(x)$ and $x = I(x')$ exactly as we require. Suitable $\delta$ exists because the representations

(51) $\qquad E \to \mathrm{End}_{\mathbb{Q}} E \qquad\qquad E \to \mathrm{End}_{\mathbb{Q}}(E^*)$

(52) $\qquad x \mapsto l(x) \qquad\qquad\qquad x \mapsto l(I(x))^*$

have the same dimension and are therefore isomorphic. We can then set $\gamma = \delta^{-1}$ so that equations (49) become equivalent to equations (50).

To ensure that $(M, b)$ can be expressed as $\mathbb{Q} \otimes_{\mathbb{Z}} (M_0, b_0)$ we can, if necessary, multiply $\gamma$ and $\delta$ by an integer.

**2.3. Proof that $M$ is Simple.** If $M' \subset M$ is a subrepresentation then $\begin{pmatrix} r(x_i) & 0 \\ 0 & 1 - r(x_i)^* \end{pmatrix}$ maps $\pi_i M'$ to itself for $i = 1$ and for $i = 2$ so in each case $\pi_i M' = V_i \oplus W_i$ where $V_i \subset E = L_i^{\oplus d}$ is invariant under right multiplication by $x_i$ and $W_i \subset E^* = (L_i^*)^{\oplus d}$ is invariant under the action of $1 - r(x_i)^*$.

The maps

$$\begin{pmatrix} 1 & \gamma \\ \delta & -1 \end{pmatrix} : V_2 \oplus W_2 \to V_1 \oplus W_1$$

$$\begin{pmatrix} 1 & -\epsilon\gamma^* \\ -\epsilon\delta^* & -1 \end{pmatrix} : V_1 \oplus W_1 \to V_2 \oplus W_2$$

imply that $V_1 = V_2$ and $W_1 = W_2$. Thus $V_1$ is invariant under multiplication by both $x_1$ and $x_2$, so $V_1 = 0$ or $E$. Since $\gamma$ and $\delta$ are isomorphisms, we have $V_1 \cong W_2$ and so $M' = 0$ or $M' = M$.

**2.4. The general case $\mu \geq 2$.** If $\mu \geq 2$ one can perform a very similar construction. One asserts that $\pi_i M = E \oplus E^*$ for $1 \leq i \leq \mu$ and defines $\rho(s) : M \to M$ by

$$\rho(s)_{ij} = \begin{cases} \begin{pmatrix} r(x_i) & 0 \\ 0 & 1 - r(x_i)^* \end{pmatrix} & \text{if } i = j \\ \begin{pmatrix} 1 & \gamma \\ \delta & -1 \end{pmatrix} & \text{if } i \neq j. \end{cases}$$

Here, $x_1, \cdots, x_\mu$ is a set of generators for $E$ over $\mathbb{Q}$ and $\gamma$ and $\delta$ are defined exactly as above.

REMARK 12.5. As in remark 12.3 above, one may add integers to the generators $x_1$ and $x_2$ to obtain for any $(E, I)$ an infinite family of integral representations $\{M_{i,E}\}_{i \in \mathbb{N}}$ and forms $b_{i,E} : M_{i,E} \to M_{i,E}^*$ such that, for each $i$, $(M_0, b_0) = (M_{i,E}, b_{i,E})$ satisfies conditions i), ii) and iii) of proposition 12.2.

## 3. Computation of $C_{2q-1}(F_\mu)$ up to Isomorphism

We are finally in a position to prove theorem A which states that if $\mu \geq 2$ and $q > 1$ then $C_{2q-1}(F_\mu)$ is isomorphic to a countable direct sum

$$(53) \quad C_{2q-1}(F_\mu) \cong \mathbb{Z}^{\oplus \infty} \oplus \left(\frac{\mathbb{Z}}{2\mathbb{Z}}\right)^{\oplus \infty} \oplus \left(\frac{\mathbb{Z}}{4\mathbb{Z}}\right)^{\oplus \infty} \oplus \left(\frac{\mathbb{Z}}{8\mathbb{Z}}\right)^{\oplus \infty}.$$

We refer to Fuchs [30] for general results on infinitely generated abelian groups.

We start, as usual, with the identification $C_{2q-1}(F_\mu) \cong W^{(-1)^q}(P_\mu\text{-}\mathbb{Z})$, assuming $q > 2$. The case $q = 2$ of (53) follows directly from the other cases because $C_3(F_\mu)$ is an index $2^\mu$ subgroup of $C_7(F_\mu)$.

Since $P_\mu$ is finitely generated, $W^\epsilon(P_\mu\text{-}\mathbb{Z})$ is countable. Now the kernel of the natural map

$$(54) \quad W^\epsilon(P_\mu\text{-}\mathbb{Z}) \to W^\epsilon(P_\mu\text{-}\mathbb{C}^-) \cong \mathbb{Z}^{\oplus \infty}$$

is 8-torsion and is therefore a direct sum of cyclic groups [30, Theorem 17.1]. The image of (54) is a subgroup of a free abelian group and hence is free abelian itself [30, Theorem 14.5]. Thus $C_{2q-1}(F_\mu)$ is a direct sum of cyclic groups of orders 2, 4, 8 and $\infty$. We must show that there are infinitely many summands of each order.

By lemma 11.1 and corollary 5.4 we have

$$(55) \quad W^\epsilon(P_\mu\text{-}\mathbb{Z}) \subset W^\epsilon(P_\mu\text{-}\mathbb{Q}) \cong \bigoplus_M W_M^\epsilon(P_\mu\text{-}\mathbb{Q}).$$

To each finite-dimensional division algebra $E$ with involution let us associate a family $(M_{i,E}, b_{i,E})_{i \in \mathbb{N}}$ of integral representations as in remark 12.5. Composing (55) with the Morita isomorphisms

$$\Theta_{\mathbb{Q} \otimes (M_{i,E}, b_{i,E})} : W^\epsilon_{\mathbb{Q} \otimes M_{i,E}}(P_\mu\text{-}\mathbb{Q}) \to W^1(E)$$

the element $[M_{i,E}] \in W^\epsilon(P_\mu\text{-}\mathbb{Z})$ is mapped to $\langle 1 \rangle \in W^1(E)$.

It will suffice for our purposes to consider commutative fields $E = K$ with *trivial* involution. By a theorem of A.Pfister, the order of $\langle 1 \rangle$ in $W^1(K)$ is twice the level of $K$ (e.g. Milnor and Husemoller [70, p75] or Scharlau [93, pp71-73]):

DEFINITION 12.6. Let $K$ be any (commutative) field. If $-1$ is a sum of squares in $K$ then the *level* of $K$ is the smallest integer $s$ such that $-1$ is a sum of $s$ squares. If $-1$ is not a sum of squares, so that $K$ is formally real, let $s = \infty$.

## 3. COMPUTATION OF $C_{2q-1}(F_\mu)$ UP TO ISOMORPHISM

Pfister showed further that the level of any field is either infinity or a power of 2 (cf theorem 7.7 above). In the case of an algebraic number field, caveat 11.8 confirms that order($\langle 1 \rangle$) $= 2s$ and implies moreover that $s$ is 1, 2 or 4 or $\infty$ (compare Lam [47, p299]).

Let $T_K \subset W^1(K)$ denote the additive subgroup of $W^1(K)$ generated by $\langle 1 \rangle$. We have $|T_K| = 2s = 2, 4, 8$ or $\infty$. If $|T_K| = 2, 4$ or $8$ then it is easy to check that $T_K$ is a pure subgroup of $W^1(K)$:

$$T_K \cap mW^1(K) = mT_K \text{ for all } m \in \mathbb{Z}.$$

It follows that $T_K$ is a direct summand of $W^1(K)$ [30, Proposition 27.1]. On the other hand, if $T_K$ is infinite then any signature is a split surjection $\sigma : W^1(K) \to \mathbb{Z}; \langle 1 \rangle \mapsto 1$ so, again, $T_K$ is a direct summand. The preimage

$$\bigoplus_{i,K} \Theta^{-1}_{\mathbb{Q}\otimes(M_{i,K}, b_{i,K})}(T_K) \subset W^\epsilon(P_\mu\text{-}\mathbb{Z}).$$

generated by the $[M_{i,K}, b_{i,K}]$ must therefore be a direct summand.

It remains to exhibit number fields $K$ of level 1, 2, 4 and $\infty$.

1) If $K = \mathbb{Q}(\sqrt{-1})$ then $-1$ is a square so $|T_K| = 2$ and $\left(\frac{\mathbb{Z}}{2\mathbb{Z}}\right)^{\oplus \infty}$ is a summand of $C_{2q-1}(F_\mu)$.

2) If $K = \mathbb{Q}(\sqrt{-3})$ then it is easy to check that $-1$ is not square, but is a sum of two squares:

$$-1 = \left(\frac{1}{2}(1+\sqrt{-3})\right)^2 + \left(\frac{1}{2}(1-\sqrt{-3})\right)^2.$$

Thus $|T_K| = 4$ and $\left(\frac{\mathbb{Z}}{4\mathbb{Z}}\right)^{\oplus \infty}$ is a summand of of $C_{2q-1}(F_\mu)$.

4) In $K = \mathbb{Q}(\sqrt{-7})$, we have $-1 = (\sqrt{-7})^2 + 2^2 + 1^2 + 1^2$. Let us show that $-1$ is not a sum of two squares. Let $\omega = \frac{1}{2}(1+\sqrt{-7})$ and let $\mathbb{Q}(\sqrt{-7})_\omega$ denote the completion of $\mathbb{Q}(\sqrt{-7})$ at the dyadic prime ideal $(w) \triangleleft \mathbb{Z}[\omega]$. If $-1$ is a sum of two squares in $\mathbb{Q}(\sqrt{-7})_\omega$ then there exists a non-trivial equation $a^2 + b^2 + c^2 = 0$ where $a, b, c$ are elements of the valuation ring $\mathbb{Z}[\omega]_\omega$ not all of which lie in the maximal ideal $(w) \triangleleft \mathbb{Z}[\omega]_\omega$. Working modulo $\omega^2$ we obtain a non-trivial equation

$$a^2 + b^2 + c^2 = 0 \quad \text{with } a, b, c \in \frac{\mathbb{Z}[\omega]_\omega}{(\omega^2)} \cong \frac{\mathbb{Z}}{4\mathbb{Z}}.$$

Only 0 and 1 are squares in $\frac{\mathbb{Z}}{4\mathbb{Z}}$ so we have reached a contradiction.

Thus $|T_K| = 8$ and $\left(\frac{\mathbb{Z}}{8\mathbb{Z}}\right)^{\oplus \infty}$ is a summand of $C_{2q-1}(F_\mu)$.

$\infty$) The field $\mathbb{Q}$ is formally real; $-1$ is not a sum of squares. Thus $T_\mathbb{Q}$ is infinite and $\mathbb{Z}^{\oplus \infty}$ is a summand of $C_{2q-1}(F_\mu)$.

EXAMPLE 12.7. Setting $E = \mathbb{Q}(\sqrt{-7})$ with trivial involution and $\epsilon = (-1)^q$, the proof of proposition 12.2 implies that the following integral representation $(\mathbb{Z}^8, \rho, \phi)$ has order 8 in $W^\epsilon(P_2\text{–}\mathbb{Z})$:

$$\rho(s) = \frac{\begin{pmatrix} 1 & 0 & 0 & 0 & 1 & 0 & 0 & 1 \\ 0 & 1 & 0 & 0 & 0 & 1 & 1 & 0 \\ 0 & 0 & 0 & 0 & 0 & 1 & -1 & 0 \\ 0 & 0 & 0 & 0 & 1 & 0 & 0 & -1 \\ 1 & 0 & 0 & -\epsilon & 0 & -7 & 0 & 0 \\ 0 & 1 & -\epsilon & 0 & 1 & 0 & 0 & 0 \\ 0 & -\epsilon & -1 & 0 & 0 & 0 & 1 & -1 \\ -\epsilon & 0 & 0 & -1 & 0 & 0 & 7 & 1 \end{pmatrix}}{},$$

$$\phi = \frac{\begin{pmatrix} 0 & 0 & 1 & 0 & 0 & 0 & 0 & 0 \\ 0 & 0 & 0 & 1 & 0 & 0 & 0 & 0 \\ \epsilon & 0 & 0 & 0 & 0 & 0 & 0 & 0 \\ 0 & \epsilon & 0 & 0 & 0 & 0 & 0 & 0 \\ 0 & 0 & 0 & 0 & 0 & 0 & 1 & 0 \\ 0 & 0 & 0 & 0 & 0 & 0 & 0 & 1 \\ 0 & 0 & 0 & 0 & \epsilon & 0 & 0 & 0 \\ 0 & 0 & 0 & 0 & 0 & \epsilon & 0 & 0 \end{pmatrix}}{}.$$

Any $F_\mu$-link (in dimension $2q - 1 > 1$) which has the corresponding Seifert matrix

$$\lambda = \phi\rho(s) = \frac{\begin{pmatrix} 0 & 0 & 0 & 0 & 0 & 1 & -1 & 0 \\ 0 & 0 & 0 & 0 & 1 & 0 & 0 & -1 \\ \epsilon & 0 & 0 & 0 & \epsilon & 0 & 0 & \epsilon \\ 0 & \epsilon & 0 & 0 & 0 & \epsilon & \epsilon & 0 \\ 0 & -\epsilon & -1 & 0 & 0 & 0 & 1 & -1 \\ -\epsilon & 0 & 0 & -1 & 0 & 0 & 7 & 1 \\ \epsilon & 0 & 0 & -1 & 0 & -7\epsilon & 0 & 0 \\ 0 & \epsilon & -1 & 0 & \epsilon & 0 & 0 & 0 \end{pmatrix}}{}$$

is therefore of order 8 in $C_{2q-1}(F_\mu)$.

# APPENDIX I

# Primitive Element Theorems

The aim of this appendix is to prove the following:

THEOREM I.1. *Suppose $k$ is a commutative field of characteristic zero and $E$ is a division $k$-algebra which is finite-dimensional over $k$. Then there exist elements $\alpha, \beta \in E$ such that $E$ is generated as a $k$-algebra by $\alpha$ and $\beta$.*

Suppose $E \subset F$ are division rings. If $F$ is finitely generated as a left $E$-module and $S \subset F$ is any subset then the ring $E\langle S\rangle$ generated in $F$ by $S$ over $E$ (i.e. the intersection of subrings of $F$ containing $E$ and $S$) is a division ring. Indeed, any non-zero $\gamma \in E\langle S\rangle$ must satisfy some equation

$$a_0 + a_1\gamma + \cdots + a_m\gamma^m = 0$$

with each $a_i \in E$ and $a_0 \neq 0$ so we have

$$\gamma^{-1} = -a_0^{-1}(a_1 + a_2\gamma + \cdots + a_m\gamma^{m-1}) \in E\langle S\rangle.$$

To prove theorem I.1 we therefore need only show that $\alpha$ and $\beta$ generate $E$ as a division ring over $k$.

DEFINITION I.2. We shall use the notation $\mathcal{I}(F/E)$ to denote the set of intermediate division rings

$$\mathcal{I}(F/E) = \{G \mid E \subset G \subset F\}.$$

We shall assume the following theorems:

THEOREM I.3. (e.g. [**50**, p243]). *Suppose $L$ is a commutative field of finite-dimension over a subfield $K$. There exists an element $\alpha \in L$ such that $K(\alpha) = L$ if and only if $\mathcal{I}(L/K)$ is finite. If $L/K$ is separable then $\mathcal{I}(L/K)$ is finite.*

THEOREM I.4. ([**18**, p110]). *Suppose $E \subset F$ are division rings and $E$ is infinite. If $\mathcal{I}(F/E)$ is finite then there exists $\alpha \in F$ such that $F = E(\alpha)$.*

THEOREM I.5 (Centralizer Theorem). (e.g. [**93**, p42] or [**20**, p42]) *Let $A$ be a simple algebra with center $K$ such that $\dim_K A < \infty$ and let $B \subset A$ be a simple sub-$K$-algebra. Then*

$$\dim_K A = \dim_K B \dim_K(Z_A(B))$$

*where*

$$Z_A(B) = \{a \in A \mid ab = ba \; \forall b \in B\}$$

*is the centralizer of $B$ in $A$.*

PROOF OF THEOREM I.1. Let $K$ be the center of $E$ and let $L$ be a maximal commutative subfield of $E$, so that $Z_E(L) = L$. Now $L$ is separable over $k$ so by theorem I.3, $L = k(\alpha)$ for some $\alpha \in L$ and the set $\mathcal{I}(L/k)$ of fields intermediate between $L$ and $k$ is finite. It follows from the centralizer theorem I.5 that
$$Z_E(Z_E(C)) = C$$
for all $C \in \mathcal{I}(E/K)$ and hence that there is a bijection
$$Z_E(\_) : \mathcal{I}(L/K) \to \mathcal{I}(E/L).$$
Therefore $\mathcal{I}(E/L)$ is finite and, by theorem I.4, there exists $\beta \in E$ such that $E = L(\beta)$ whence $E = k(\alpha, \beta)$ as required. □

# APPENDIX II

# Hermitian Categories

In this appendix we prove that a duality preserving functor $(F, \Phi, \eta)$ between hermitian categories is an equivalence if and only if the underlying functor $F$ is an equivalence.

## 1. Functors

We first recall some definitions of category theory. Let $\mathcal{C}$ and $\mathcal{D}$ be categories and let $F : \mathcal{C} \to \mathcal{D}$ be a functor. $F$ is said to be *faithful* if for all objects $M$, $M'$ in $\mathcal{C}$, the induced map

(56) $$F : \text{Hom}_{\mathcal{C}}(M, M') \to \text{Hom}_{\mathcal{D}}(F(M), F(M'))$$

is injective. If this induced map (56) is surjective for all $M$ and $M'$ one says that $F$ is *full*. $F$ is an *equivalence of categories* if there exists a functor $F' : \mathcal{D} \to \mathcal{C}$ and natural isomorphisms $\alpha : F'F \to \text{id}_{\mathcal{C}}$ and $\beta : FF' \to \text{id}_{\mathcal{D}}$.

LEMMA II.1. *An equivalence of categories is faithful and full.*

In fact, $F$ is an equivalence of categories if and only if $F$ is faithful and full and every object in $\mathcal{D}$ is isomorphic to $F(M)$ for some object $M \in \mathcal{C}$ (see Bass [5, p4]).

LEMMA II.2. *If $F : \mathcal{C} \to \mathcal{D}$ is an equivalence of categories with inverse $F'$ then one can choose natural isomorphisms $\alpha : F'F \to \text{id}_{\mathcal{C}}$ and $\beta : FF' \to \text{id}_{\mathcal{D}}$ such that $F'(\beta_N) = \alpha_{F'(N)}$ for all objects $N \in \mathcal{D}$ and $F(\alpha_M) = \beta_{F(M)}$ for all objects $M \in \mathcal{C}$.*

PROOF. Fix a natural isomorphism $\alpha : F'F \to \text{id}_{\mathcal{C}}$. For each $N \in \mathcal{D}$ define $\beta_N : FF'(N) \to N$ by $F'(\beta_N) = \alpha_{F'(N)}$. We must check that $\beta$ is a natural isomorphism and that $\beta_{F(M)} = F(\alpha_M)$.

Suppose $N$ and $N'$ are objects of $\mathcal{D}$ and $f \in \text{Hom}_{\mathcal{D}}(N, N')$. Using the naturality of $\alpha$ we have $F'(f)F'(\beta_N) = F'(f)\alpha_{F'(N)} = \alpha_{F'(N')}F'FF'(f) : F'FF'(N) \to F'(N')$. Since $F'$ is faithful by lemma II.1, $f\beta_N = \beta_{N'}FF'(f)$ so $\beta$ is a natural transformation.

Since $F'$ is full and $F(\beta_N) = \alpha_{F'(N)}$ has an inverse for each $N$, $\beta_N$ is also invertible. Thus $\beta$ is a natural isomorphism.

Finally, since $\alpha$ is natural there is a commutative square

$$\begin{array}{ccc} F'FF'F(M) & \xrightarrow{\alpha_{F'F(M)}} & F'F(M) \\ {\scriptstyle F'F(\alpha_M)}\downarrow & & \downarrow{\scriptstyle \alpha_M} \\ F'F(M) & \xrightarrow{\alpha_M} & M \end{array}$$

so $F'F(\alpha_M) = \alpha_{F'F(M)}$ and therefore $F'(\beta_{F(M)}) = \alpha_{F'F(M)} = F'F(\alpha_M)$. Since $F'$ is faithful, $\beta_{F(M)} = F(\alpha_M)$ as required. □

## 2. Duality Preserving Functors

In definition 3.19 the notion was introduced of a duality preserving functor $(F, \Phi, \eta) : \mathcal{C} \to \mathcal{D}$ between hermitian categories.

DEFINITION II.3. Composition of duality preserving functors is given by

$$(F', \Phi', \eta') \circ (F, \Phi, \eta) = (F' \circ F, \Phi'F'(\Phi), \eta'\eta)$$

with $(\Phi'F'(\Phi))_M = \Phi'_{F(M)}F'(\Phi_M)$.

The identity morphism on a hermitian category $\mathcal{C}$ is $(\mathrm{id}_{\mathcal{C}}, \{\mathrm{id}_M^*\}_{M\in\mathcal{C}}, 1)$, which will be abbreviated $\mathrm{id}_{\mathcal{C}}$.

DEFINITION II.4. Suppose $(F, \Phi, \eta), (F', \Phi', \eta) : \mathcal{C} \to \mathcal{D}$ are duality preserving functors. A natural transformation $\alpha : (F, \Phi, \eta) \to (F', \Phi', \eta)$ is a natural transformation $\alpha : F \to F'$ such that

$$(57) \qquad \alpha_M^* \Phi'_M \alpha_{M^*} = \Phi_M : F(M^*) \to F(M)^*$$

for all $M \in \mathcal{C}$.

LEMMA II.5. If $(F', \Phi', \eta) : \mathcal{C} \to \mathcal{D}$ is a duality preserving functor and $\alpha : F \to F'$ is a natural transformation then equation (57) defines a pullback $\Phi$ such that $(F, \Phi, \eta)$ is a duality preserving functor and $\alpha : (F, \Phi, \eta) \to (F', \Phi', \eta)$ is a natural transformation.

PROOF. It suffices to check that $\Phi_M^* i_{F(M)} = \eta \Phi_{M^*} F(i_M)$. Indeed,

$$\begin{aligned} \Phi_M^* i_{F(M)} &= (\alpha_M^* \Phi'_M \alpha_{M^*})^* i_{F(M)} \\ &= \alpha_{M^*}^* (\Phi'_M)^* \alpha_M^{**} i_{F(M)} \\ &= \alpha_{M^*}^* (\Phi'_M)^* i_{F'(M)} \alpha_M \quad \text{by naturality of } i \\ &= \eta \alpha_{M^*}^* \Phi'_{M^*} F'(i_M) \alpha_M \\ &= \eta \alpha_{M^*}^* \Phi'_{M^*} \alpha_{M^{**}} F(i_M) \\ &= \eta \Phi_{M^*} F(i_M). \end{aligned}$$
□

DEFINITION II.6. A duality preserving functor $(F, \Phi, \eta) : \mathcal{C} \to \mathcal{D}$ is said to be an equivalence of hermitian categories if there exists $(F', \Phi', \eta) : \mathcal{D} \to \mathcal{C}$ such that $(F', \Phi', \eta) \circ (F, \Phi, \eta)$ is naturally isomorphic to $\mathrm{id}_{\mathcal{C}} = (\mathrm{id}_{\mathcal{C}}, \{\mathrm{id}_M^*\}_{M\in\mathcal{C}}, 1)$ and $(F, \Phi, \eta) \circ (F', \Phi', \eta)$ is naturally isomorphic to $\mathrm{id}_{\mathcal{D}}$.

## 2. DUALITY PRESERVING FUNCTORS

Hermitian categories $\mathcal{C}$ and $\mathcal{D}$ are said to be $\eta$-equivalent if there is an equivalence $(F, \Phi, \eta) : \mathcal{C} \to \mathcal{D}$.

PROPOSITION II.7. *A duality preserving functor $(F, \Phi, \eta) : \mathcal{C} \to \mathcal{D}$ is an equivalence of hermitian categories if and only if $F : \mathcal{C} \to \mathcal{D}$ is an equivalence of categories.*

PROOF. The 'only if' part follows directly from the definitions. Conversely, suppose $(F, \Phi, \eta)$ is a duality preserving functor, and $F$ is an equivalence. As in lemma II.2, there exists $F' : \mathcal{D} \to \mathcal{C}$ and natural isomorphisms $\alpha : F'F \to \mathrm{id}_\mathcal{C}$ and $\beta : FF' \to \mathrm{id}_\mathcal{D}$ such that $F'(\beta_N) = \alpha_{F'(N)}$ for all objects $N \in \mathcal{D}$ and $F(\alpha_M) = \beta_{F(M)}$ for all objects $M \in \mathcal{C}$.

We aim to define $\Phi' : F'(\_^*) \to F'(\_)^*$ such that $(F', \Phi', \eta)$ is a duality preserving functor and $\alpha : (F', \Phi', \eta) \circ (F, \Phi, \eta) \to \mathrm{id}_\mathcal{C}$ and $\beta : (F, \Phi, \eta) \circ (F', \Phi', \eta) \to \mathrm{id}_\mathcal{D}$ are natural isomorphisms. $\Phi'$ must satisfy the following equations for all $M \in \mathcal{C}$ and $N \in \mathcal{D}$:

(58) $\quad (\Phi'_N)^* i_{F'(N)} = \eta \Phi'_{N^*} F'(i_N) : F'(N) \to F'(N^*)^*;$

(59) $\quad \Phi'_{F(M)} F'(\Phi_M) = \alpha_M^* \alpha_{M^*} : F'F(M^*) \to F'F(M)^*;$

(60) $\quad \Phi_{F'(N)} F(\Phi'_N) = \beta_N^* \beta_{N^*} : FF'(N^*) \to FF'(N)^*.$

Since $F$ is faithful, equation (60) serves as a definition of $\Phi'_N$.

*Proof of equation* (58): Since $F$ is faithful, it suffices to check that

$$F((\Phi'_N)^*) F(i_{F'(N)}) = \eta F(\Phi'_{N^*}) FF'(i_N).$$

Composing on the left by $\Phi_{F'(N^*)}$ and applying equation (60) to the right hand side we aim to show

$$\Phi_{F'(N^*)} F((\Phi'_N)^*) F(i_{F'(N)}) = \eta \beta_{N^*}^* \beta_{N^{**}} FF'(i_N).$$

Indeed,

$$\Phi_{F'(N^*)} F((\Phi'_N)^*) F(i_{F'(N)}) = F(\Phi'_N)^* \Phi_{F'(N)^*} F(i_{F'(N)}) \quad \text{by naturality of } \Phi$$

$$= \eta F(\Phi'_N)^* \Phi_{F'(N)}^* i_{FF'(N)} \quad \text{by equation (20)}$$

$$= \eta (\beta_N^* \beta_{N^*})^* i_{FF'(N)} \quad \text{by equation (60)}$$

$$= \eta \beta_{N^*}^* \beta_N^{**} i_{FF'(N)}$$

$$= \eta \beta_{N^*}^* i_N \beta_N \quad \text{by naturality of } i$$

$$= \eta \beta_{N^*}^* \beta_{N^{**}} FF'(i_N) \quad \text{by naturality of } \beta.$$

*Proof of equation* (59): Applying $F$ and composing on the left by $\Phi_{F'F(M)}$ it suffices to check that

$$\Phi_{F'F(M)} F(\Phi'_{F(M)}) FF'(\Phi_M) = \Phi_{F'F(M)} F(\alpha_M^*) F(\alpha_{M^*}).$$

By equation (60) the left hand side is $\beta^*_{F(M)}\beta_{F(M)^*}FF'(\Phi_M)$ while

$$\Phi_{F'F(M)}F(\alpha^*_M)F(\alpha_{M^*}) = F(\alpha_M)^*\Phi_M F(\alpha_{M^*}) \quad \text{by naturality of } \Phi$$
$$= \beta^*_{F(M)}\Phi_M \beta_{F(M^*)}$$
$$= \beta^*_{F(M)}\beta_{F(M)^*}F'F(\Phi_M) \quad \text{by naturality of } \beta. \quad \square$$

# Bibliography

[1] A. A. Albert, *Structure of Algebras*, Amer. Math. Soc. Colloq. Publ., vol. 24, American Mathematical Society, New York, 1939.

[2] M. Auslander, I. Reiten, and S. O. Smalø, *Representation Theory of Artin Algebras*, Cambridge Stud. Adv. Math., 36, Cambridge University Press, 1995.

[3] H.-J. Bartels, *Invarianten hermitescher Formen über Schiefkörpern*, Math. Ann. **215** (1975), 269–288.

[4] _____, *Zur Klassifikation Schiefhermitescher Formen über Zahlkörpern*, Math. Ann. **219** (1976), no. 1, 13–19.

[5] H. Bass, *Algebraic K-theory*, W. A. Benjamin Inc, New York-Amsterdam, 1968.

[6] D. J. Benson, *Representations and cohomology.I. Basic representation theory of finite groups and associative algebras*, Cambridge Stud. Adv. Math., 30, Cambridge University Press, 1995.

[7] R. C. Blanchfield, *Intersection theory of manifolds with operators with applications to knot theory*, Ann. of Math. (2) **65** (1957), 340–356.

[8] J. Bocknak, M. Coste, and M.-F. Roy, *Real Algebraic Geometry*, Springer, Berlin, 1998.

[9] N. Bourbaki, *Éléments de Mathématique, Book 2 Ch.8*, Hermann, Paris, 1958.

[10] S. E. Cappell and J. L. Shaneson, *Topological knots and knot cobordism*, Topology **12** (1973), 33–40.

[11] _____, *The codimension two placement problem, and homology equivalent manifolds*, Ann. of Math. (2) **99** (1974), 277–348.

[12] _____, *Link cobordism*, Comment. Math. Helv. **55** (1980), 20–49.

[13] A. J. Casson and C. McA. Gordon, *On slice knots in dimension three*, Algebraic and geometric topology, Proc. Sympos. Pure Math., no. XXXII, Part 2, American Mathematical Society, Providence RI, 1978, pp. 39–53.

[14] _____, *Cobordism of classical knots*, À la recherche de la topologie perdue, Progr. Math., 62, Birkhäuser Boston, Boston, MA, 1986, pp. 181–199.

[15] T. D. Cochran and K. E. Orr, *Not all links are concordant to boundary links*, Bull. Amer. Math. Soc. (N.S.) **23** (1990), no. 1, 99–106.

[16] _____, *Not all links are concordant to boundary links*, Ann. of Math. (2) **138** (1993), 519–554.

[17] T. D. Cochran, K. E. Orr, and P. Teichner, *Knot concordance, Whitney towers and $L^2$-signatures*, e-print math.GT/9908117, 1999.

[18] P. M. Cohn, *Skew Fields, Theory of General Division Rings*, Encyclopedia Math. Appl., 57, Cambridge University Press, 1995.

[19] C. W. Curtis and I. Reiner, *Methods of Representation Theory. Vol I. With Applications to Finite Groups and Orders*, John Wiley & Sons, New York, 1981.

[20] P. K. Draxl, *Skew Fields*, London Math. Soc. Lecture Note Ser., 81, Cambridge University Press, 1983.

[21] D. Dubois and G. Efroymson, *A dimension theorem for real primes*, Canad. J. Math. **26** (1974), no. 1, 108–114.

[22] J. Duval, *Forme de Blanchfield et cobordisme d'entrelacs bords*, Comment. Math. Helv. **61** (1986), no. 4, 617–635.

[23] M. Farber, *The classification of simple knots*, Uspekhi Mat. Nauk. **38** (1983), no. 5, 59–106, Russian Math. Surveys 38:5 (1983) 63-117.

[24] _____, *Hermitian forms on link modules*, Comment. Math. Helv. **66** (1991), no. 2, 189–236.

[25] _____, *Noncommutative rational functions and boundary links*, Math.Ann. **293** (1992), no. 3, 543–568.

[26] _____, *Stable-homotopy and homology invariants of boundary links*, Trans. Amer. Math. Soc. **334** (1992), no. 1, 455–477.

[27] M. Farber and P. Vogel, *The Cohn localization of the free group ring*, Math. Proc. Cambridge Philos. Soc. **111** (1992), no. 3, 433–443.

[28] E. Formanek, *Generating the ring of matrix invariants*, Ring theory (Antwerp 1985), Lecture Notes in Math., 1197, Springer, Berlin, 1986, pp. 73–82.

[29] R. H. Fox and J. Milnor, *Singularities of 2-spheres in 4-space and cobordism of knots*, Osaka J. Math. **3** (1966), 257–267.

[30] L. Fuchs, *Infinite Abelian Groups. Volume I*, Pure and Applied Mathematics, 36, Academic Press, New York and London, 1970.

[31] P. Gilmer, *Slice knots in $S^3$*, Quart. J. Math. Oxford Ser. (2) **34** (1983), no. 135, 305–322.

[32] _____, *Classical knot and link concordance*, Comment. Math. Helv. **68** (1993), no. 1, 1–19.

[33] M. A. Gutiérrez, *Boundary links and an unlinking theorem*, Trans. Amer. Math. Soc. **171** (1972), 491–499.

[34] I. Hambleton and I. Madsen, *On the computation of the projective surgery obstruction groups*, $K$-theory **7** (1993), no. 6, 537–574.

[35] J. A. Hillman, *Alexander ideals of links*, Lecture Notes in Math., 895, Springer, Berlin-New York, 1981.

[36] J. A. Hillman, *Algebraic Invariants of Links*, World Scientific Publishing, to appear.

[37] M. W. Hirsch, *Differential Topology*, Grad. Texts in Math., No. 33, Springer, 1976.

[38] C. Kearton, *Blanchfield duality and simple knots*, Trans. Amer. Math. Soc. **202** (1975), 141–160.

[39] _____, *Cobordism of knots and Blanchfield duality*, J. London Math. Soc. (2) **10** (1975), no. 4, 406–408.

[40] M. Kervaire and C. Weber, *A survey of multidimensional knots*, Proc. 1977 Plans Conf. Knot Theory, Lecture Notes in Math., 685, Springer, 1978, pp. 61–134.

[41] M. A. Kervaire, *Les noeuds de dimensions supérieures*, Bull. Soc. Math. France **93** (1965), 225–271.

[42] _____, *Knot cobordism in codimension two*, Manifolds–Amsterdam 1970, Lecture Notes in Math., 197, Springer, Berlin, 1971, pp. 83–105.

[43] Paul Kirk and Charles Livingston, *Twisted Alexander invariants, Reidemeister torsion and Casson-Gordon invariants*, Topology **38** (1999), no. 3, 635–661.

[44] M.-A. Knus, *Quadratic and Hermitian Forms over Rings*, Grundlehren der Mathematischen Wissenschaften, 294, Springer, Berlin, 1991.

[45] K. H. Ko, *Seifert matrices and boundary link cobordisms*, Trans. Amer. Math. Soc. **299** (1987), no. 2, 657–681.

[46] _____, *A Seifert-matrix interpretation of Cappell and Shaneson's approach to link cobordisms*, Math. Proc. Cambridge Philos. Soc. **106** (1989), 531–545.

[47] T. Y. Lam, *The Algebraic Theory of Quadratic Forms*, W.A.Benjamin, Reading, Massachusetts, 1973.

[48] _____, *An introduction to real algebra*, Rocky Mountain J. Math. **14** (1984), no. 4, 767–814.

[49] W. Landherr, *Äquivalenz Hermitescher Formen über einem beliebigen algebraischen Zahlkörper*, Abh. Math. Sem. Univ. Hamburg **11** (1936), 245–248.
[50] S. Lang, *Algebra*, 3rd ed., Addison-Wesley, 1993.
[51] L. Le Bruyn and C. Procesi, *Semisimple representations of quivers*, Trans. Amer. Math. Soc. **317** (1990), no. 2, 585–598.
[52] J.-Y. Le Dimet, *Cobordisme d'enlacements de disques*, Bull. Soc. Math. France **116** (1988), ii+92, Mémoire no. 32.
[53] C. F. Letsche, *An obstruction to slicing knots using the eta invariant*, Math. Proc. Cambridge Philos. Soc. **128** (2000), no. 2, 301–319.
[54] J. Levine, *Unknotting spheres in codimension two*, Topology **4** (1965), 9–16.
[55] _____, *Invariants of knot cobordism*, Invent. Math. **8** (1969), 98–110, Addendum, 8:355.
[56] _____, *Knot cobordism groups in codimension two*, Comment. Math. Helv. **44** (1969), 229–244.
[57] _____, *Knot modules I*, Trans. Amer. Math. Soc. **229** (1977), 1–50.
[58] _____, *Signature invariants of homology bordism with applications to links*, Knots 90 (Osaka 1990), de Gruyter, Berlin, 1992, pp. 395–406.
[59] _____, *Link invariants via the eta invariant*, Comment. Math. Helv. **69** (1994), no. 1, 82–119.
[60] J. Levine and K. E. Orr, *A survey of applications of surgery to knot and link theory*, Surveys on surgery theory, Vol 1, Ann. of Math. Stud., 145, Princeton University Press, Princeton, NJ, 2000, pp. 345–364.
[61] D. W. Lewis, *A note on Hermitian and quadratic forms*, Bull. London Math. Soc. **11** (1979), no. 3, 265–267.
[62] _____, *The isometry classification of Hermitian forms over division algebras*, Linear Algebra Appl. **43** (1982), 245–272.
[63] _____, *New improved exact sequences of Witt groups*, J. Algebra **74** (1982), no. 1, 206–210.
[64] _____, *Quaternionic skew-Hermitian forms over a number field*, J. Algebra **74** (1982), no. 1, 232–240.
[65] W. B. R. Lickorish, *An introduction to knot theory*, Grad. Texts in Math., 175, Springer, New York, 1997.
[66] D. Luna, *Slices étales*, Bull. Soc. Math. France Mémoirs **33** (1973), 81–105.
[67] _____, *Adhérances d'orbite et invariants*, Invent. Math. **29** (1975), no. 3, 231–238.
[68] J. W. Milnor, *Infinite cyclic coverings*, Conference on the Topology of Manifolds, Prindle, Weber & Schmidt, Boston, Mass, 1968, pp. 115–133.
[69] _____, *On isometries of inner product spaces*, Invent. Math. **8** (1969), 83–97.
[70] J. W. Milnor and D. Husemoller, *Symmetric Bilinear Forms*, Springer, 1973.
[71] W. Mio, *On boundary-link cobordism*, Math. Proc. Cambridge Philos. Soc. **101** (1987), 259–266.
[72] D. Mumford, J. Fogarty, and F. Kirwan, *Geometric Invariant Theory*, 3rd ed., Springer, Berlin, 1994.
[73] W. D. Neumann, *Equivariant Witt Rings*, Bonner Math. Schriften, 100, Universität Bonn, Mathematisches Institut, 1977.
[74] S. P. Novikov, *On manifolds with free Abelian fundamental groups and their applications*, Izvestiya Akademii Nauk SSSR. Seriya Matematicheskaya (1966), 207–246, Amer. Math. Soc. Transl. Ser. 2, 71 (1968), 1-42.
[75] O. T. O'Meara, *Introduction to quadratic forms*, Die Grundlehren der mathematischen Wissenschaften, 117, Academic Press, New York; Springer, Berlin, 1963.
[76] W. Pardon, *Local surgery and applications to the theory of quadratic forms*, Bull. Amer. Math. Soc. **82** (1976), no. 1, 131–133.
[77] _____, *Local surgery and the exact sequence of a localization for Wall groups*, Mem. Amer. Math. Soc. **12** (1977), no. 196, iv+171.

[78] C. Procesi, *The invariant theory of $n \times n$ matrices*, Adv. Math. **19** (1976), 306–381.

[79] _____, *A formal inverse to the Cayley-Hamilton theorem*, J. Algebra **107** (1987), 63–74.

[80] H.-G. Quebbemann, W. Scharlau, and M. Schulte, *Quadratic and Hermitian forms in additive and abelian categories*, J. Algebra **59** (1979), no. 2, 264–289.

[81] A. A. Ranicki, *The algebraic theory of surgery I. Foundations*, Proc. London Math. Soc. (3) **40** (1980), no. 1, 87–192.

[82] _____, *The algebraic theory of surgery II. Applications to topology*, Proc. London Math. Soc. (3) **40** (1980), no. 2, 193–283.

[83] _____, *Exact Sequences in the Algebraic Theory of Surgery*, Princeton University Press, New Jersey; University of Tokyo Press, Tokyo, 1981.

[84] _____, *High-dimensional Knot Theory*, Springer, Berlin, 1998.

[85] V. Retakh, C. Reutenauer, and A. Vaintrob, *Noncommutative rational functions and Farber's invariants of boundary links*, Differential topology, infinite-dimensional Lie algebras, and applications, Amer. Math. Soc. Transl., Ser. 2, 194, American Mathematical Society, Providence, RI, 1999, pp. 237–246.

[86] L. Ribes and P. Zalesskii, *Profinite Groups*, Ergeb. Math. Grenzgeb. (3), 40, Springer, Berlin, 2000.

[87] D. Rolfsen, *Knots and Links*, Mathematics Lecture Series, 7, Publish or Perish, Inc, Berkeley, California, 1990, Corrected reprint of the 1976 original.

[88] J. Rosenberg, *Algebraic K-theory and its applications*, Grad. Texts in Math., 147, Springer, New York, 1994.

[89] N. Sato, *Algebraic invariants of boundary links*, Trans. Amer. Math. Soc. **265** (1981), no. 2, 359–374.

[90] _____, *Free coverings and modules of boundary links*, Trans. Amer. Math. Soc. **264** (1981), no. 2, 499–505.

[91] _____, *Alexander modules*, Proc. Amer. Math. Soc. **91** (1984), no. 1, 159–162.

[92] W. Scharlau, *Induction theorems and the structure of the Witt group*, Invent. Math. **11** (1970), 37–44.

[93] _____, *Quadratic and Hermitian forms*, Grundlehren Math. Wiss., 270, Springer, Berlin, 1985.

[94] J.-P. Serre, *Linear Representations of Finite Groups*, Grad. Texts in Math., 42, Springer, New York-Heidelberg, 1977.

[95] J. R. Smith, *Complements of codimension-two submanifolds - III - cobordism theory*, Pacific J. Math. **94** (1981), no. 2, 423–484.

[96] N. Smythe, *Boundary links*, Topology Seminar, Wisconsin, 1965 (R.H. Bing and R.J. Bean, eds.), Ann. of Math. Stud., 60, Princeton University Press, Princeton, N.J., 1966, pp. 69–72.

[97] N. W. Stoltzfus, *Unraveling the integral knot concordance group*, Mem. Amer. Math. Soc., 192, vol. 12, Issue 1, American Mathematical Society, Providence, Rhode Island, 1977.

[98] A. G. Tristram, *Some cobordism invariants for links*, Proceedings of the Cambridge Philosophical Society **66** (1969), 251–264.

[99] P. Vogel, *Localisation in algebraic L-theory*, Proc. 1979 Siegen Topology Conf., Lecture Notes in Math., 788, Springer, 1980, pp. 482–495.

[100] _____, *On the obstruction group in homology surgery*, Publ. Math. I.H.E.S. **55** (1982), 165–206.

[101] C. T. C. Wall, *Surgery on Compact Manifolds*, 2nd ed., Math. Surveys Monogr., vol. 69, American Mathematical Society, 1999, (1st edition published 1970).

Dept of Mathematics, UC Riverside,
California 92521, USA.
des@sheiham.com

# Index

$(F, \Phi, \eta)$, 35
$(M, \phi)$, 32
$(M, \rho)$, 29
$(R-A)$-Proj, 29
$(R-\mathbb{Q})_\mathbb{Z}$-Proj, 88
$(R-\mathbb{Q}/\mathbb{Z})$-Proj, 89
$A$-Proj, 31
$A^o$, 32
$A_\mu$, 6
$B(n, \mu)$, 5
$C(n, \mu)$, 3
$\mathbb{C}^-, \mathbb{C}^+$, 18, 33
$C_n(F_\mu)$, 6
$\mathrm{End}(M)$, 22
$\mathrm{GL}(\alpha)$, 53
$G^{\epsilon,\mu}(A)$, 9, 14, **38**
$\Gamma_n(\mathbb{Z}[\pi] \to \mathbb{Z})$, 12, 13
$H^\epsilon(A)$, 32
$H^\epsilon(\mathcal{C})$, 34
$\mathcal{I}(F/E)$, 99
$K_0(A)$, 31
$K_0(R-A)$, 32
$K_0(\mathcal{C})$, 31
$l(x), r(x)$, 92
$L_n(A)$, 11, 33
$\overline{\mathcal{M}}^s(\mathcal{C}, \epsilon)$, 48
$\mathcal{M}(R) = \mathcal{M}(R-\mathbb{C})$, 51
$\overline{\mathcal{M}}(R), \mathcal{M}^s(R), \overline{\mathcal{M}}^s(R)$, 51
$\mathcal{M}(R, \alpha), \overline{\mathcal{M}}(R, \alpha)$, 20, 53, 54
$\mathcal{O}, \mathcal{O}_K$, 73
$P_\mu$, 19, **31**, 38
$\sigma_{M,b}(L, \theta)$, 26
$W^\epsilon(A)$, 33
$W^\epsilon(R-\mathbb{Q}/\mathbb{Z})$, 89
$W^\epsilon(\mathcal{C}), W^\epsilon(R-A)$, 36
$W^\epsilon_\mathbb{Z}(R-\mathbb{Q})$, 88

Addition of knots and $F_\mu$-links, 6
Admissible subobject, 35
Alexander polynomial, 10, 18
Algebra
    Artin, 63
    cyclic, 82

    quaternion, 22, **82**

Blanchfield form, 10, 15, 19
Blanchfield-Duval form, 15, 17
Boundary link, 5

Character, 20, **63–65**
    independence of, 64
Cobordism
    of boundary links, 5
    of $F_\mu$-links, 6
    of links, 3
    of Seifert surfaces, 5, 6

Devissage, 24, **47–50**
    hermitian, 48
Dimension vector, 20, **30**
Discriminant, 22, **85**
Duality functor, 20, **34**
Duality preserving functor, **35**, 101
    composition of, 102
    natural transformation between, 102

$\epsilon$-hermitian form, 32, 34
$\epsilon$-self-dual, 34

$F_\mu$-link, 5
Frobenius reciprocity, 59, 61

Grothendieck group, 31

Hasse-Minskowski map, 82
Hasse-Witt invariant, 22, **86**
Hermitian
    category, 34
        equivalence of, 102
    form, 32

Involution, 20, 32
    non-standard, 22, **83**
    of the first kind, 83
    of the second kind, **83**, 84
    standard, 83
Isotopy, 2

Jordan-Hölder theorem, 24, **47**

Knot, 2

Level of a field, 96
Lewis $\theta$-invariant, 22, **87**
Link, 2
  boundary, 5
  cobordism, 3
  $F_\mu$-, 5
  split, 7, 20
Localization exact sequence, 12, 15, 27, 88
Luna stratum, 20, 54

Metabolic, 9, 14, **36**, 37
Metabolizer (=Lagrangian), 33, **36**
Morita equivalence, 25, **41–45**

Null-cobordant, 3

Ordered field, 59

Path ring, 19, **30**
Pfister's theorem, 59

Quiver, 19, **29**
  complete, 20

Real
  algebraic set, 54
  closed field, 60
  closure of a field, 60
  nullstellensatz, 55
  radical, 55
  ring (=Formally real ring), 55
  variety, 55
Representation, 20, **29**
  algebraic, 72
  algebraically integral, 21, **73**
  conjugate, 72
  induced, 67
  integral, 27
  of a quiver, 30
  restriction of, 68
  self-dual, 20
  simple (=irreducible), 20, **64**
  type, 54

Seifert
  form, 8, 9, 14–15, **37**
  surface, 3, 5
Self-dual, 34
Signature, 18, 23, 26
Slice, 3, 9
  boundary-, 5, 8
Sublagrangian, 36
Surgery, **10**, 7–15
  homology, 10, 13
  on a Seifert surface, 8–10
Sylvester's theorem, 23

Variety of representations, 20, **51**

Witt group
  of a hermitian category, 35–37
  of a ring with involution, 32
  of Blanchfield forms, 10
  of Seifert forms, 9

## Editorial Information

To be published in the *Memoirs*, a paper must be correct, new, nontrivial, and significant. Further, it must be well written and of interest to a substantial number of mathematicians. Piecemeal results, such as an inconclusive step toward an unproved major theorem or a minor variation on a known result, are in general not acceptable for publication. Papers appearing in *Memoirs* are generally longer than those appearing in *Transactions*, which shares the same editorial committee.

As of June 1, 2003, the backlog for this journal was approximately 3 volumes. This estimate is the result of dividing the number of manuscripts for this journal in the Providence office that have not yet gone to the printer on the above date by the average number of monographs per volume over the previous twelve months, reduced by the number of volumes published in four months (the time necessary for preparing a volume for the printer). (There are 6 volumes per year, each containing at least 4 numbers.)

A Consent to Publish and Copyright Agreement is required before a paper will be published in the *Memoirs*. After a paper is accepted for publication, the Providence office will send a Consent to Publish and Copyright Agreement to all authors of the paper. By submitting a paper to the *Memoirs*, authors certify that the results have not been submitted to nor are they under consideration for publication by another journal, conference proceedings, or similar publication.

## Information for Authors

*Memoirs* are printed from camera copy fully prepared by the author. This means that the finished book will look exactly like the copy submitted.

The paper must contain a *descriptive title* and an *abstract* that summarizes the article in language suitable for workers in the general field (algebra, analysis, etc.). The *descriptive title* should be short, but informative; useless or vague phrases such as "some remarks about" or "concerning" should be avoided. The *abstract* should be at least one complete sentence, and at most 300 words. Included with the footnotes to the paper should be the 2000 *Mathematics Subject Classification* representing the primary and secondary subjects of the article. The classifications are accessible from www.ams.org/msc/. The list of classifications is also available in print starting with the 1999 annual index of *Mathematical Reviews*. The Mathematics Subject Classification footnote may be followed by a list of *key words and phrases* describing the subject matter of the article and taken from it. Journal abbreviations used in bibliographies are listed in the latest *Mathematical Reviews* annual index. The series abbreviations are also accessible from www.ams.org/publications/. To help in preparing and verifying references, the AMS offers MR Lookup, a Reference Tool for Linking, at www.ams.org/mrlookup/. When the manuscript is submitted, authors should supply the editor with electronic addresses if available. These will be printed after the postal address at the end of the article.

**Electronically prepared manuscripts.** The AMS encourages electronically prepared manuscripts, with a strong preference for $\mathcal{AMS}$-LaTeX. To this end, the Society has prepared $\mathcal{AMS}$-LaTeX author packages for each AMS publication. Author packages include instructions for preparing electronic manuscripts, the *AMS Author Handbook*, samples, and a style file that generates the particular design specifications of that publication series. Though $\mathcal{AMS}$-LaTeX is the highly preferred format of TeX, author packages are also available in $\mathcal{AMS}$-TeX.

Authors may retrieve an author package from e-MATH starting from **www.ams.org/tex/** or via FTP to **ftp.ams.org** (login as **anonymous**, enter username as password, and type **cd pub/author-info**). The *AMS Author Handbook* and the *Instruction Manual* are available in PDF format following the author packages link from **www.ams.org/tex/**. The author package can be obtained free of charge by sending email to **pub@ams.org** (Internet) or from the Publication Division, American Mathematical Society, 201 Charles St., Providence, RI 02904, USA. When requesting an author package, please specify $\mathcal{A}_\mathcal{M}\mathcal{S}$-LaTeX or $\mathcal{A}_\mathcal{M}\mathcal{S}$-TeX, Macintosh or IBM (3.5) format, and the publication in which your paper will appear. Please be sure to include your complete mailing address.

**Sending electronic files.** After acceptance, the source file(s) should be sent to the Providence office (this includes any TeX source file, any graphics files, and the DVI or PostScript file).

Before sending the source file, be sure you have proofread your paper carefully. The files you send must be the EXACT files used to generate the proof copy that was accepted for publication. For all publications, authors are required to send a printed copy of their paper, which exactly matches the copy approved for publication, along with any graphics that will appear in the paper.

TeX files may be submitted by email, FTP, or on diskette. The DVI file(s) and PostScript files should be submitted only by FTP or on diskette unless they are encoded properly to submit through email. (DVI files are binary and PostScript files tend to be very large.)

Electronically prepared manuscripts can be sent via email to **pub-submit@ams.org** (Internet). The subject line of the message should include the publication code to identify it as a Memoir. TeX source files, DVI files, and PostScript files can be transferred over the Internet by FTP to the Internet node **e-math.ams.org** (130.44.1.100).

**Electronic graphics.** Comprehensive instructions on preparing graphics are available at **www.ams.org/jourhtml/graphics.html**. A few of the major requirements are given here.

Submit files for graphics as EPS (Encapsulated PostScript) files. This includes graphics originated via a graphics application as well as scanned photographs or other computer-generated images. If this is not possible, TIFF files are acceptable as long as they can be opened in Adobe Photoshop or Illustrator. No matter what method was used to produce the graphic, it is necessary to provide a paper copy to the AMS.

Authors using graphics packages for the creation of electronic art should also avoid the use of any lines thinner than 0.5 points in width. Many graphics packages allow the user to specify a "hairline" for a very thin line. Hairlines often look acceptable when proofed on a typical laser printer. However, when produced on a high-resolution laser imagesetter, hairlines become nearly invisible and will be lost entirely in the final printing process.

Screens should be set to values between 15% and 85%. Screens which fall outside of this range are too light or too dark to print correctly. Variations of screens within a graphic should be no less than 10%.

**Inquiries.** Any inquiries concerning a paper that has been accepted for publication should be sent directly to the Electronic Prepress Department, American Mathematical Society, 201 Charles St., Providence, RI 02904, USA.

## Editors

This journal is designed particularly for long research papers, normally at least 80 pages in length, and groups of cognate papers in pure and applied mathematics. Papers intended for publication in the *Memoirs* should be addressed to one of the following editors. In principle the Memoirs welcomes electronic submissions, and some of the editors, those whose names appear below with an asterisk (*), have indicated that they prefer them. However, editors reserve the right to request hard copies after papers have been submitted electronically. Authors are advised to make preliminary email inquiries to editors about whether they are likely to be able to handle submissions in a particular electronic form.

**Algebraic geometry** to DAN ABRAMOVICH, Department of Mathematics, Boston University, 111 Cummington St., Boston, MA 02215; email: abramovic@bu.edu

**Algebraic topology and cohomology of groups** to STEWART PRIDDY, Department of Mathematics, Northwestern University, 2033 Sheridan Road, Evanston, IL 60208-2730; email: priddy@math.nwu.edu

**Combinatorics and Lie theory** to SERGEY FOMIN, Department of Mathematics, University of Michigan, Ann Arbor, Michigan 48109-1109; email: fomin@umich.edu

**Complex analysis and complex geometry** to DUONG H. PHONG, Department of Mathematics, Columbia University, 2990 Broadway, New York, NY 10027-0029; email: phong@math.columbia.edu

*__Differential geometry and global analysis__ to LISA C. JEFFREY, Department of Mathematics, University of Toronto, 100 St. George St., Toronto, ON Canada M5S 3G3; email: jeffrey@math.toronto.edu

**Dynamical systems and ergodic theory** to ROBERT F. WILLIAMS, Department of Mathematics, University of Texas, Austin, Texas 78712-1082; email: bob@math.utexas.edu

*__Geometric analysis__ to TOBIAS COLDING, Courant Institute, New York University, 251 Mercer St., New York, NY 10012; email: colding@cims.nyu.edu

**Harmonic analysis** to ALEXANDER NAGEL, Department of Mathematics, University of Wisconsin, 480 Lincoln Drive, Madison, WI 53706-1313; email: nagel@math.wisc.edu

**Harmonic analysis, representation theory, and Lie theory** to ROBERT J. STANTON, Department of Mathematics, The Ohio State University, 231 West 18th Avenue, Columbus, OH 43210-1174; email: stanton@math.ohio-state.edu

**Number theory** to HAROLD G. DIAMOND, Department of Mathematics, University of Illinois, 1409 W. Green St., Urbana, IL 61801-2917; email: diamond@math.uiuc.edu

*__Ordinary differential equations, and applied mathematics__ to PETER W. BATES, Department of Mathematics, Michigan State University, East Lansing, MI 48824-1027; email: peter@math.msu.edu

*__Partial differential equations__ to PATRICIA E. BAUMAN, Department of Mathematics, Purdue University, West Lafayette, IN 47907-1395' email: bauman@math.purdue.edu

*__Probability and statistics__ to KRZYSZTOF BURDZY, Department of Mathematics, University of Washington, Box 354350, Seattle, Washington 98195-4350; email: burdzy@math.washington.edu

*__Real analysis and partial differential equations__ to DANIEL TATARU, Department of Mathematics, University of California, Berkeley, Berkeley, CA 94720; email: tataru@math.berkeley.edu

**All other communications to the editors** should be addressed to the Managing Editor, WILLIAM BECKNER, Department of Mathematics, University of Texas, Austin, TX 78712-1082; email: beckner@math.utexas.edu.

# Titles in This Series

787 **Michael Cwikel, Per G. Nilsson, and Gideon Schechtman,** Interpolation of weighted Banach lattices/A characterization of relatively decomposable Banach lattices, 2003

786 **Arnd Scheel,** Radially symmetric patterns of reaction-diffusion systems, 2003

785 **R. R. Bruner and J. P. C. Greenlees,** The connective K-theory of finite groups, 2003

784 **Desmond Sheiham,** Invariants of boundary link cobordism, 2003

783 **Ethan Akin, Mike Hurley, and Judy A. Kennedy,** Dynamics of topologically generic homeomorphisms, 2003

782 **Masaaki Furusawa and Joseph A. Shalika,** On central critical values of the degree four $L$-functions for GSp(4): The Fundamental Lemma, 2003

781 **Marcin Bownik,** Anisotropic Hardy spaces and wavelets, 2003

780 **S. Marmi and D. Sauzin,** Quasianalytic monogenic solutions of a cohomological equation, 2003

779 **Hansjörg Geiges,** $h$-principles and flexibility in geometry, 2003

778 **David B. Massey,** Numerical control over complex analytic singularities, 2003

777 **Robert Lauter,** Pseudodifferential analysis on conformally compact spaces, 2003

776 **U. Haagerup, H. P. Rosenthal, and F. A. Sukochev,** Banach embedding properties of non-commutative $L^p$-spaces, 2003

775 **P. Lochak, J.-P. Marco, and D. Sauzin,** On the splitting of invariant manifolds in multidimensional near-integrable Hamiltonian systems, 2003

774 **Kai A. Behrend,** Derived $\ell$-adic categories for algebraic stacks, 2003

773 **Robert M. Guralnick, Peter Müller, and Jan Saxl,** The rational function analogue of a question of Schur and exceptionality of permutation representations, 2003

772 **Katrina Barron,** The moduli space of $N = 1$ superspheres with tubes and the sewing operation, 2003

771 **Shigenori Matsumoto,** Affine flows on 3-manifolds, 2003

770 **W. N. Everitt and L. Markus,** Elliptic partial differential operators and symplectic algebra, 2003

769 **Jie Wu,** Homotopy theory of the suspensions of the projective plane, 2003

768 **R. Höpfner and E. Löcherbach,** Limit theorems for null recurrent Markov processes, 2003

767 **Po Hu,** $S$-modules in the category of schemes, 2003

766 **Su Gao and Alexander S. Kechris,** On the classification of Polish metric spaces up to isometry, 2003

765 **Robert Bieri and Ross Geoghegan,** Connectivity properties of group actions on non-positively curved spaces, 2003

764 **J. Spandaw,** Noether-Lefschetz problems for degeneracy loci, 2003

763 **Yasuyuki Kachi and Eiichi Sato,** Segre's reflexivity and an inductive characterization os hyperquadrics, 2002

762 **Leiba Rodman, Ilya M. Spitkovsky, and Hugo Woerdeman,** Abstract band method via factorization, positive and band extensions of multivariable almost periodic matrix functions, and spectral estimation, 2002

761 **Oliver Druet and Emmanuel Hebey,** The $AB$ program in geometric analysis : Sharp Sobolev inequalities and related problems, 2002

760 **Markus Banagl,** Extending intersection homology type invariants to non-Witt spaces, 2002

759 **Donald M. Davis,** From representation theory to homotopy groups, 2002

758 **Alan Forrest, John Hunton, and Johannes Kellendonk,** Topological invariants for projection method patterns, 2002

## TITLES IN THIS SERIES

757 **Douglas Bowman,** $q$-difference operators, orthogonal polynomials, and symmetric expansions, 2002

756 **José Ignacio Cogolludo-Agustín,** Topological invariants of the complement to arrangements of rational plane curves, 2002

755 **M. A. Mandell and J. P. May,** Equivariant orthogonal spectra and $S$-modules, 2002

754 **Edward L. Green, Idun Reiten, and Øyvind Solberg,** Dualities on generalized Koszul algebras, 2002

753 **Daniel Panazzolo,** Desingularization of nilpotent singularities in families of planar vector fields, 2002

752 **Linus Kramer,** Homogeneous spaces, Tits buildings, and isoparametric hypersurfaces, 2002

751 **Bruce Allison, Georgia Benkart, and Yun Gao,** Lie algebras graded by the root systems $BC_r$, $r \geq 2$, 2002

750 **Masaki Izumi and Hideki Kosaki,** Kac algebras arising from composition of subfactors: General theory and classification, 2002

749 **Nanhua Xi,** The based ring of two-sided cells of affine Weyl groups of type $\widetilde{A}_{n-1}$, 2002

748 **Jürgen Ritter and Alfred Weiss,** The lifted root number conjecture and Iwasawa theory, 2002

747 **Armand Borel, Robert Friedman, and John W. Morgan,** Almost commuting elements in compact Lie groups, 2002

746 **Peter Niemann,** Some generalized Kac-Moody algebras with known root multiplicities, 2002

745 **Mikhail A. Lifshits and Werner Linde,** Approximation and entropy numbers of Volterra operators with application to Brownian motion, 2002

744 **Roger Chalkley,** Basic global relative invariants for homogeneous linear differential equations, 2002

743 **Heng Sun,** Spectral decomposition of a covering of $GL(r)$: the Borel case, 2002

742 **J. E. Gilbert, Y. S. Han, J. A. Hogan, J. D. Lakey, D. Weiland, and G. Weiss,** Smooth molecular functions and singular integral operators, 2002

741 **Francisco Santos,** Triangulations of oriented matroids, 2002

740 **Rick Durrett,** Mutual invadability implies coexistence in spatial models, 2002

739 **Georgios K. Alexopoulos,** Sub-Laplacians with drift on Lie groups of polynomial volume growth, 2002

738 **Yasuro Gon,** Generalized Whittaker functions on $SU(2,2)$ with respect to the Siegel parabolic subgroup, 2002

737 **Arjen Doelman, Robert A. Gardner, and Tasso J. Kaper,** A stability index analysis of 1-D patterns of the Gray-Scott model, 2002

736 **Wojciech Chachólski and Jérôme Scherer,** Homotopy theory of diagrams, 2002

735 **Martina Brück, Xi Du, Joonsang Park, and Chuu-Lian Terng,** The submanifold geometries associated to Grassmannian systems, 2002

734 **Michel Van den Bergh,** Blowing up of non-commutative smooth surfaces, 2001

733 **Milé Krajčevski,** Tilings of the plane, hyperbolic groups and small cancellation conditions, 2001

732 **Jan O. Kleppe, Juan C. Migliore, Rosa Miró-Roig, Uwe Nagel, and Chris Peterson,** Gorenstein liaison, complete intersection liaison invariants and unobstructedness, 2001

For a complete list of titles in this series, visit the
AMS Bookstore at **www.ams.org/bookstore/**.